Kinetics of Chemical Processes

BUTTERWORTH–HEINEMANN SERIES IN CHEMICAL ENGINEERING

SERIES EDITOR
HOWARD BRENNER
Massachusetts Institute of Technology

ADVISORY EDITORS
ANDREAS ACRIVOS
The City College of CUNY
JAMES E. BAILEY
California Institute of Technology
MANFRED MORARI
California Institute of Technology
E. BRUCE NAUMAN
Rensselaer Polytechnic Institute
J.R.A. PEARSON
Schlumberger Cambridge Research
ROBERT K. PRUD'HOMME
Princeton University

SERIES TITLES

Bubble Wake Dynamics in Liquids and Liquid-Solid Suspensions *Liang-Shih Fan and Katsumi Tsuchiya*

Chemical Process Equipment: Selection and Design *Stanley M. Walas*

Chemical Process Structures and Information Flows *Richard S.H.Mah*

Computational Methods for Process Simulations *W. Fred Ramirez*

Constitutive Equations for Polymer Melts and Solutions *Ronald G. Larson*

Fluidization Engineering, 2nd ed. *Octave Levenspiel*

Fundamental Process Control *David M. Prett and Carlos E. Garcia*

Gas-Liquid-Solid Fluidization Engineering *Liang-Shih Fan*

Gas Separation by Adsorption Processes *Ralph T. Yang*

Granular Filtration of Aerosols and Hydrosols *Chi Tien*

Heterogeneous Reactor Design *Hong H. Lee*

Introductory Systems Analysis for Process Engineers *E. Bruce Nauman*

Microhydrodynamics: Principles and Selected Applications *Sangtae Kim and Seppo J. Karrila*

Modelling With Differential Equations *Stanley M. Walas*

Molecular Thermodynamics of Nonideal Fluids *Lloyd L. Lee*

Phase Equilibria in Chemical Engineering *Stanley M. Walas*

Physicochemical Hydrodynamics: An Introduction *Ronald F. Probstein*
Principles and Practice of Slurry Flow *C.A. Shook and M.C. Roco*
Transport Processes in Chemically Reacting Flow Systems *Daniel E. Rosner*
Viscous Flows: The Practical Use of Theory *Stuart W. Churchill*

REPRINT TITLES

Advanced Process Control *W. Harmon Ray*
Applied Statistical Mechanics *Thomas M. Reed and Keith E. Gubbins*
Elementary Chemical Reactor Analysis *Rutherford Aris*
Kinetics of Chemical Processes *Michel Boudart*
Reaction Kinetics for Chemical Engineers *Stanley M. Walas*

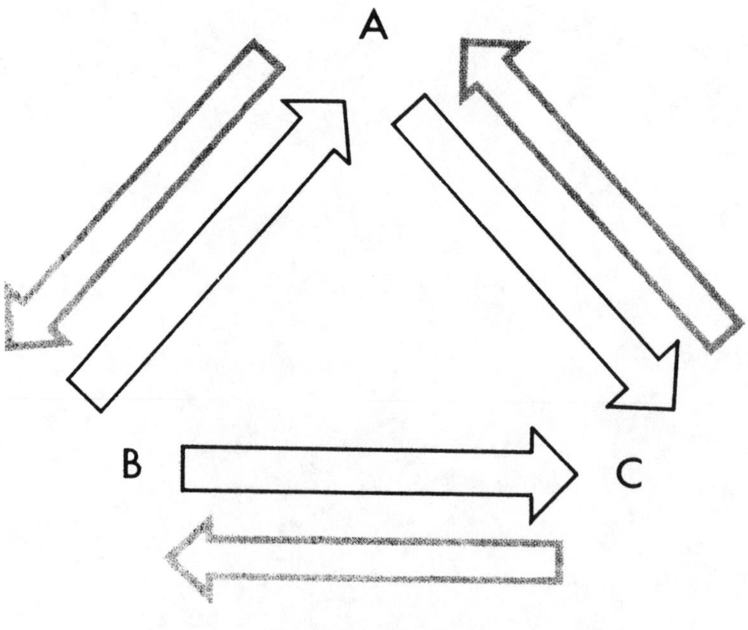
A
B
C

Kinetics of Chemical Processes

MICHEL BOUDART
Professor of Chemical Engineering and Chemistry
Stanford University

BUTTERWORTH–HEINEMANN
Boston London Oxford Singapore Sydney Toronto Wellington

This Butterworth–Heinemann edition is an unabridged reprint of the book originally published by Prentice-Hall in 1968.

Library of Congress Cataloging-in-Publication Data

Boudart, Michel.
Kinetics of chemical processes / Michel Boudart.
p. cm.—(Butterworth–Heinemann series in chemical engineering)
Includes bibliographical references and index.
ISBN 0–7506–9006–2 (paperback)
1. Chemical reaction, Rate of. I. Title. II. Series.
QD501.B7818 1991
541.3′9—dc20 91–8227
CIP

British Library Cataloguing in Publication Data
Boudart, Michel
Kinetics of chemical processes.
1. Chemical reactions. Kinetics
I. Title
541.394

ISBN 0–7506–9006–2

Butterworth–Heinemann
80 Montvale Avenue
Stoneham, MA 02180

10 9 8 7 6 5 4 3 2 1

Printed in the United States of America

Contents

Preface and Erratum

This book is essentially 25 years old, as it was written in the mid-sixties. Its claim for originality resides in the chosen organization of the subject matter and in an attempt at quantitative description of the rate of chemical processes based on reaction mechanisms but not dominated by the latter. In a more recent book (*Kinetics of Heterogeneous Catalytic Reactions,* by M. Boudart and G. Djéga-Mariadassou, Princeton University Press, Princeton, NJ, 1984), we embellish these thoughts in the following manner.

> Kinetics is not only a tool of pure or applied research, but also a very satisfying avocation. Reaction mechanisms come and go, and their ephemeral existence is often disconcerting. By contrast, the results of good chemical kinetics remain unchanged, whatever may be the future revisions of their underlying mechanism. The chemist in chemical dynamics is frequently accused by the engineer of *explaining* everything without ever *predicting* anything. Yet the primary goal of kinetics is to *describe* the chemical transformation. A good description possesses permanent value, which remains the foundation of any future explanation or prediction. To contribute something of lasting value is a normal human aspiration. The kinetics of the pioneers of heterogeneous catalysis has retained all of its value. Today, after a temporary eclipse, heterogeneous catalytic kinetics has rejoined the vanguard of surface science and catalysis science, and has become once more a respectable endeavor for chemists and chemical engineers.

Thus, if the old book shows its age, it is primarily in the choice of symbols and abbreviations that antedate the current recommendations of IUPAC. Of course, this does not mean that nothing has happened in chemical kinetics since 1968. Enormous advances have been made in the past 20 years in the kinetics of elementary steps (Chapter 2), especially from the viewpoint of the detailed manner by which they take place from selected energy states of reactants to determined energy states of products. Yet this informative invasion into the private lives of reacting molecules does not detract from the classical treatment of transition state theory as presented in Chapter 2.

Another advance in chemical kinetics has also taken place in the field of the free radical chain reactions with applications to polymerization, steam cracking and combustion. This advance takes advantage of our ability to compute reaction rates of hundreds of simultaneous elementary steps involved in very large networks of reactions (Chapter 5). A similar development is currently taking place in heterogeneous catalytic kinetics following the bold efforts of Jim Dumesic and co-workers at the University of Wisconsin. In this revolution brought about by a growing base of kinetic data and by the broad availability of computers, the traditional procedure has reversed itself. In the past, macroscopic observations of rates were analyzed to yield mechanisms and rate constants. In the future, rates will be calculated from rate constants of elementary steps, tabulated, approximated or calculated. But to understand what really goes on, the principles presented in the *Kinetics of Chemical Processes* will be as helpful as they were in the past, perhaps even more so.

Over the years, no serious error in the book has been reported to me. But to avoid being accused of perfection, I shall close by reporting one erratum.

Erratum

The rate equation (5.3.13) as found in the paper of Bateman et al., *Disc. Far. Soc.*, ***10,*** 250 (1951) has been re-examined in the light of the correct definitions of rates given in the first chapter of the Notes. The rate of termination of $\mathbf{R_1OO}$ should be:

$$r_{t_{\mathbf{R_1OO}}} = -\frac{d(\mathbf{R_1OO})}{dt} = 2k'_{11}(\mathbf{R_1OO})^2 + k'_{12}(\mathbf{R_1OO})(\mathbf{R_2OO})$$

And the rate of termination of $\mathbf{R_2OO}$ should be:

$$r_{t_{\mathbf{R_2OO}}} = -\frac{d(\mathbf{R_2OO})}{dt} = 2k'_{22}(\mathbf{R_2OO})^2 + k'_{12}(\mathbf{R_1OO})(\mathbf{R_2OO})$$

Hence the total rate of termination of active centers is:

$$-\frac{d(\mathbf{R_1OO})}{dt} + \frac{d(\mathbf{R_2OO})}{dt} = 2k'_{11}(\mathbf{R_1OO})^2 + 2k'_{12}(\mathbf{R_1OO})(\mathbf{R_2OO}) + 2k'_{22}(\mathbf{R_2OO})^2$$

Therefore the correct form of the rate equation (5.3.13) is

$$r_0 = r_i^{1/2}\frac{\rho_1(R_1H)^2 + 2(R_1H)(R_2H) + \rho_2(R_2H)^2}{[2\alpha_1(R_1H)^2 + 2\beta(R_1H)(R_2H) + 2\alpha_2(R_2H)^2]^{1/2}}$$

Equation 5.3.14 as found in the paper of Tsepalov, *Zhur. Fiz. Khim.*, ***35***, 1086, 1443, 1691 (1962) should be $\alpha_1 = 2k'_{11}/k_{12}^2$ and $\alpha_2 = 2k'_{22}/k_{21}^2$. The appropriate changes should be made where necessary, i.e., substitute $2\alpha_1$ for α_1 and $2\alpha_2$ for α_2. Then

$$(\delta r/\delta z)_z = 0 = \frac{2(2\alpha_1)^{1/2} - \rho_1(2\alpha_1)^{-1/2}\beta}{2\alpha_1}.$$

This will be positive or negative according to whether $4/\rho_1 - \beta/\alpha_1$ is positive or negative, thus, R_2H is an accelerator if $4k_{21}/k_{11} > k'_{12}/k'_{11}$.

Preface to the Original Printing

Chemical kinetics is a tool in the hands of organic and inorganic chemists who search for a quantitative formulation of chemical reactivity. It also provides the essential data which permit the chemical engineer to design, operate, control and optimize the reactors of the chemical industry.

Thus, chemical kinetics has long ceased to be the exclusive preoccupation of the kineticist, whose concepts and theories are now commonly used in pure and applied chemistry. But in spite of its increasing popularity, chemical kinetics is not generally taught in a course of physical chemistry except as one chapter out of many others.

The present book attempts to develop this chapter into a full-fledged course, while presenting only essential ideas without stressing any particular type of chemistry. All it tries to do is to explain the concepts associated with the kinetic study of the chemical process A goes to B goes to C. It may seem unfortunate that compounds A, B and C have neither odor nor taste nor color. But if they did, the subject matter of this book would be chemistry and not chemical kinetics. It is hoped that this ascetic presentation of the subject will help in forging a more generally usable tool for pure and applied research. Examples have been used only to avoid giving the impression that chemical kinetics is but a branch of theoretical physics. Alas, this is not true at the time of writing.

This book does not present kinetics for the kineticist, i.e., the more advanced topics in development, including the studies of reactive scattering and the rates of energy transfer in molecular systems. But the recent vintage of many of the ideas and illustrations in every chapter should convince anyone that this is not a book about the chemical kinetics of the past generation. In fact, activity in classical chemical kinetics is so universal that any attempt to sketch the subject in less than three hundred pages requires a very eclectic approach.

In the last analysis, the choice of topics has been dictated by my own taste and experience as well as by the needs of many hundreds of chemists and chemical engineers on the campuses of Princeton, Berkeley and Stanford to whom a course now crystallized in this book has been presented over the past eight years at the undergraduate and graduate level. The material has also been tried in accelerated courses of one or two weeks at Humble Oil and Refining Company, Esso Research and Engineering Company and The Chemstrand Corporation.

To all my students in industry and at the university, I wish to express my thanks for providing the attentive criticism which has forced me to deepen, refine and correct many of my kinetic ideas.

I also wish to thank my friends and mentors: Sir Hugh Taylor who introduced me to the subject, John Fenn who convinced me that I should study it, Richard Wilhelm and Charles Wilke who encouraged me to teach it to chemical engineers. I am also much indebted to my colleague Robert J. Madix for his careful preparation of solutions to the problems given in the book.

M. BOUDART

Stanford

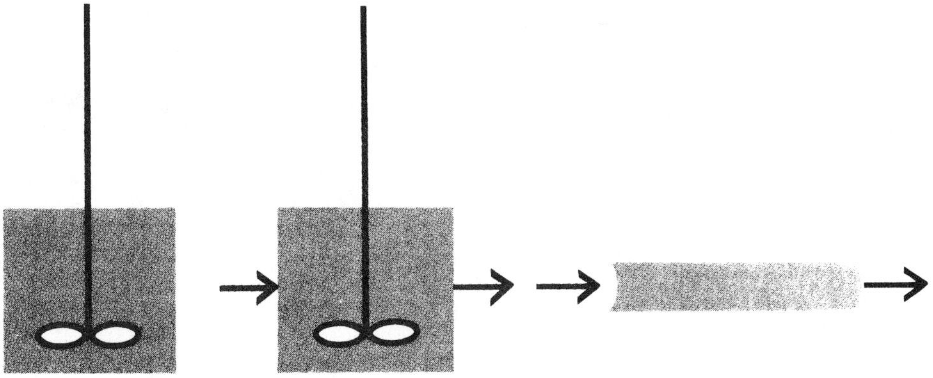

Introduction

1

1.1 The Scope of Chemical Kinetics

The evolution in time of chemically reacting systems is the object of study of chemical kinetics. More specifically, chemical kinetics concerns itself with the measurement and interpretation of reaction rates. The information provides the quantitative basis underlying all theories of chemical reactivity. Thus, chemical kinetics is a tool in the search for new knowledge of molecular behavior, and the measurement of reaction rates is a means to an end, not an end in itself.

But chemical kinetics is also an essential tool in the research and development of new processes. It serves to define the set of values of process variables that will best fulfill the requirements of practical operation. Thus chemical kinetics is a tool of considerable value to theoretical chemists and chemical engineers alike. The difference in emphasis is best seen by looking at the structure of the problems arising in the study of reaction rates.

In general, in a chemically reacting system, many simultaneous reactions take place via consecutive and parallel paths. We say that we deal with a *network of reactions*. An example of a simple triangular network is the isomerization of butenes:

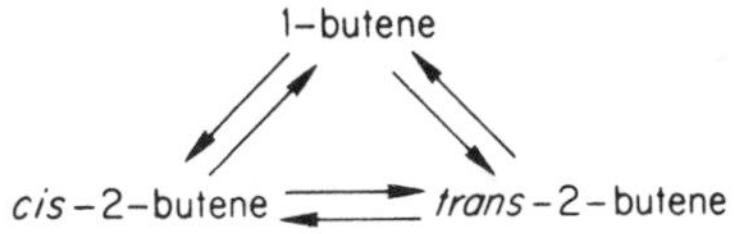

In biochemistry such networks, comprising dozens of parallel and consecutive reactions, are frequently called *pathways*.

The first task of the kineticist is the identification of the reactions and their proper arrangement in the network. A kinetic analysis of the network provides valuable clues to the solution of this problem. Ultimately, it leads to information on the rates of individual reactions.

Each reaction of the network is stoichiometrically simple in the sense that its advancement is described by a single parameter: the extent of reaction (see next section). A stoichiometrically simple reaction will be called a *single reaction* or a *reaction*, for short. The expression "simple reaction" is best avoided since, in general, a stoichiometrically simple reaction is far from simple. Indeed, in the vast majority of cases, a single reaction does not take place as written. It proceeds through a *sequence of steps* involving reactive intermediates that do not appear in the equation for reaction. In what follows, a sequence of steps will be called a sequence. The identification of these intermediates and the definition of the proper sequence are the central problems of the kinetic analysis. This is logically the second task of the kineticist but it is not the last one.

Each step of the sequence is elementary; it proceeds at the molecular level as written. It represents an irreducible molecular event. In this book, elementary steps will be called steps for short. As an example of a stoichiometrically simple reaction consider the hydrogenation of bromine:

$$H_2 + Br_2 \rightarrow 2\ HBr$$

It does not proceed as written. Rather, it proceeds in a sequence of two steps, involving hydrogen and bromine atoms that do not appear in the equation for reaction but actually exist in the reacting system in very small concentrations:

$$Br + H_2 \rightarrow HBr + H$$

$$H + Br_2 \rightarrow HBr + Br$$

The remaining third task of the kineticist is to find directly or indirectly the rate of these individual *steps*. With the help of theory, the detailed stereochemistry of each rearrangement can then be examined and knowledge about chemical reactivity accumulated systematically.

In this panoramic survey of chemical kinetics, the words "mechanism" or "model" have been excluded on purpose. "Mechanism" or "model" can mean an assumed reaction network, or a plausible sequence of steps for a given reaction, or a postulated stereochemical path during the course of an isolated step. Since methods of investigation and goals are so utterly different in the study of networks, sequences and steps, the words "mechanism" or "model" should be avoided. They have acquired the bad connotation associated with irresponsible or vain speculation, largely to describe achievements that vary widely in sophistication.

As the reacting system evolves from reactants to products, a number of *intermediates* appear, reach a certain concentration and ultimately vanish. Corresponding to the distinction between networks, reactions and steps, three different kinds of intermediates can be recognized. First there are intermediates of reactivity, concentration and lifetime comparable to those of stable reactants and products. These intermediates are the ones that appear in the reactions of the network. A typical intermediate of this first kind is formaldehyde CH_2O in the oxidation of methane (see below).

Then there are intermediates which appear in the sequence of steps for an individual reaction of the network. These are much more reactive than the former. They are usually present in very small concentrations and their lifetime is short as compared to that of initial reactants. These reactive intermediates will be called *active centers* to distinguish them from the more stable entities which will be called intermediates for short. Typical active centers are the hydrogen and bromine atoms in the reaction between hydrogen and bromine molecules.

Finally, each elementary step proceeds from reactants to products through an intermediate called a *transition state* which by definition cannot be isolated and must be considered as a species in transit. A celebrated example of a transition state is symbolized on the cover of this book or on top of the Chemistry Conference Building at Stanford University by a trigonal bipyramid. This configuration is reached for instance in the transition state of the step:

$$OH^- + C_2H_5Br \rightarrow HOC_2H_5 + Br^-$$

The transition state would then be represented as shown in Fig. 1.1.1.

Thus the study of elementary steps naturally focuses attention on transition states and the elementary step may be considered as understood if the structure of the reacting system can be defined completely at any time from the mutual approach of reactants until separation of the products through the transition state. The kinetics of steps then represents the foundation of chemical kinetics and the highest level of understanding of chemical reactivity.

But in the vast majority of cases, considerable knowledge of a theoretical or practical nature can be derived from a study of the reaction at a much less

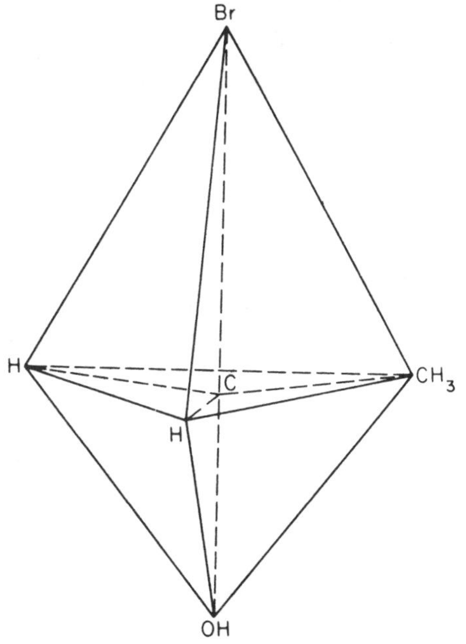

Fig. 1.1.1 The transition state of the elementary step:
$OH^- + C_2H_5Br \rightarrow OHC_2H_5 + Br^-$
The nucleophilic substituent OH^- displaces the leaving group Br^-.

fundamental level. A mere dissection of the reaction into its elementary steps will yield information on the nature and reactivity of the active centers. In the early stages of a kinetic investigation, the questions to be settled are even of a more qualitative nature. The first task is the dissection of the network into its reactions, with recognition of the stable intermediates participating in these reactions.

At what level a kinetic study will be conducted will then be dictated by the purpose of the work and the state of development of the subject. It is very important, in each case, to define the scope of the kinetic investigation and thus also its proper methodology and limitations.

As an example, consider the oxidation of methane around 700°K and at pressures close to atmospheric. The system evolves from two stable reactants,

CH_4 and O_2, to two stable products CO_2 and H_2O, through a network of four successive single reactions:

$$CH_4 + O_2 \rightarrow CH_2O + H_2O$$
$$CH_2O + O_2 \rightarrow CO + H_2O_2$$
$$CO + \tfrac{1}{2} O_2 \rightarrow CO_2$$
$$H_2O_2 \rightarrow H_2O + \tfrac{1}{2} O_2$$

The intermediates are formaldehyde CH_2O, hydrogen peroxide H_2O_2 and carbon monoxide CO. To gather enough information in support of such a network requires special methods which will be discussed in Chapter 10. Each reaction must then be dissected into its elementary steps. For instance, the oxidation of formaldehyde to carbon monoxide and hydrogen peroxide

$$CH_2O + O_2 \rightarrow CO + H_2O_2$$

takes place through the postulated sequence of two steps:

$$CHO + O_2 \rightarrow CO + HO_2$$
$$HO_2 + CH_2O \rightarrow H_2O_2 + CHO$$

involving two active centers CHO and HO_2.

The kinetic approach to the treatment of such a problem will be presented in Chapters 3 and 4. The effects due to the interaction or coupling of two reactions, such as

$$CH_4 + O_2 \rightarrow CH_2O + H_2O$$
$$CH_2O + O_2 \rightarrow CO + H_2O_2$$

through the active centers taking part in these reactions will be discussed in Chapter 5. The self-acceleration of the oxidation of methane due to formaldehyde will be explained in Chapter 6. The minimum information on the kinetics of elementary steps involved can be found in Chapter 2 while useful empirical correlations which, in this instance give apparently a very good prediction of the maximum amount of formaldehyde, will be presented in Chapter 8. Finally, as in all chemical processes, interference from gradients of concentration and temperature is always to be feared. In the oxidation of methane, as in many exothermic processes, a thermal explosion is possible. Such phenomena are outlined in Chapter 7.

Generally, a distinction between organic and inorganic reactions or homogeneous and heterogeneous processes will be avoided throughout the book, which is not concerned with any particular chemistry but only with general principles. Nevertheless, because of the many particular features of reactions at surfaces, a special chapter has been set aside for heterogeneous kinetics (Chapter 9).

In this introductory chapter, we present the definition of rates of reaction, the general properties of the mathematical function representing the rate as well as the behavior of the ideal reactors used in the measurement of reaction rates.

Problem 1.1.1

From your own experience in organic, inorganic or general chemistry, present a convincing example of a network and show its dissection into single reactions and elementary steps. List the stable intermediates and active centers and, if possible, suggest the structure of at least one transition state.

This problem should be continued as you proceed through the book and be completed only at the end of the course.

1.2 The Extent of Reaction

When a chemical reaction takes place in a system it is frequently, though not always, possible to characterize the change by a stoichiometric equation. It is convenient to write this equation in the following general way:

$$0 = \sum_i \nu_i \mathrm{A}_i \tag{1.2.1}$$

The sum is taken over all components A_i of the system. The stoichiometric coefficients ν_i are positive for products, negative for reactants and equal to zero for inert components that do not take part in the reaction. The convention of sign accounts for the inverted form of (1.2.1) which is rarely used in practice.

Of course, there are many equivalent ways to write the stoichiometric equation for a reaction. For example, in the case of ammonia synthesis, a literal application of (1.2.1) would give:

$$0 = 2\ \mathrm{NH_3} - \mathrm{N_2} - 3\ \mathrm{H_2}$$

This is a pedantic version of the accepted form with reactants on the left-hand side and products on the right-hand side:

$$\mathrm{N_2} + 3\ \mathrm{H_2} = 2\ \mathrm{NH_3}$$

which is preferred provided that the sign convention is kept in mind:

$$\nu_{N_2} = -1, \qquad \nu_{H_2} = -3, \qquad \nu_{NH_3} = 2$$

Another acceptable version is

$$\tfrac{1}{2} N_2 + \tfrac{3}{2} H_2 = NH_3$$

and the choice is purely a matter of taste and convenience. Sometimes it is expedient to write the equation in such a way that the stoichiometric coefficient of a component of interest is unity, in absolute value. However, since there is no universal convention on this point, it is always essential to write down the stoichiometric equation of a reaction before starting any thermodynamic or kinetic discussion of its behavior.

Consider now a closed system, i.e., a system which exchanges no mass with its surroundings. Initially, there are n_i^0 moles of component A_i. If the reaction described by (1.2.1) takes place, the number of moles n_i of each component A_i at any time t will be given by the relation:

$$n_i = n_i^0 + \nu_i X \qquad (1.2.2)$$

which is the expression of the *Law of Definite Proportions* and also the definition of the parameter X called the *extent of reaction.* This parameter changes in time and is a natural reaction variable. Its drawback is that it is an extensive variable, i.e., it is proportional to the mass of the system under consideration. The *fractional conversion* f does not suffer from this defect and it can readily be related to X. In general, reactants are not present initially in stoichiometric proportions. Then, the reactant present in the least amount determines the maximum possible value of the extent of reaction X_{max}. This component, called the limiting reactant (subscript l), will be totally consumed when X reaches its maximum value X_{max}:

$$0 = n_l^0 + \nu_l X_{max} \qquad (1.2.3)$$

The fractional conversion f is then simply defined as:

$$f = \frac{X}{X_{max}} \qquad (1.2.4)$$

and can be calculated from (1.2.3):

$$f = (-\nu_l)\left(\frac{X}{n_l^0}\right) = 1 - \frac{n_l}{n_l^0} \qquad (1.2.5)$$

Many different reaction variables can similarly be derived from the extent of reaction. Only two will be cited here.

Use is frequently made of the *conversion variable* x related to X by the relation:

$$x = \frac{X}{V} \tag{1.2.6}$$

where V is the volume of the system. The variable x should be used only when the volume of the system does not change during the course of the reaction.

Another reaction variable, sometimes called the *efficiency of the reaction*, suggests itself when the progress of the reaction is limited by the position of chemical equilibrium. When the thermodynamic variables are such that X cannot reach its maximum value X_{max} but will approach its equilibrium value X_e smaller than X_{max}, the efficiency of the reaction is defined as:

$$\eta = \frac{X}{X_e} \tag{1.2.7}$$

When a reaction is limited by thermodynamic equilibrium in this fashion, we shall say that the reaction is *reversible*. When X_e is equal to X_{max} for all practical purposes, the reaction will be called *irreversible*. The endpoint of the reaction characterized by $X_{max} = X_e$ can still be considered as the equilibrium point toward which the system is striving.

Problem 1.2.1

If there are several simultaneous reactions taking place in a closed system, each one is characterized by its own extent of reaction. Generalize Eq. (1.2.2) so that it will apply to a system where r different reactions take place simultaneously.

Problem 1.2.2

Apply the general equation derived in Problem 1.2.1 to the following situation. Carbon monoxide is burned with the stoichiometric amount of air. Because of the high temperature, the equilibrium

$$N_2 + O_2 \rightleftarrows 2\,NO \tag{1}$$

has to be taken into account, besides the equilibrium:

$$CO + \tfrac{1}{2}\,O_2 \rightleftarrows CO_2 \tag{2}$$

Total pressure is one atmosphere. At the temperature of interest, the equilibrium constants of reactions (1) and (2) are respectively $K_1 = 8.26 \times 10^{-3}$ and $K_2 = 7.9$. What is the equilibrium composition of the system, expressed in mole fractions?

1.3 *The Rate of Reaction*

Consider a closed system of uniform pressure, temperature and composition in which a single chemical reaction takes place, as represented by the stoichiometric equation (1.2.1). The extent of reaction X defined in (1.2.2) increases with time t. The rate of reaction is then defined as:

$$R = \frac{dX}{dt} \tag{1.3.1}$$

This quantity is either *positive* or equal to zero when for some reason the reaction is prevented from taking place or the system has reached equilibrium.

The reaction rate R is, like X, an extensive property of the system. Traditionally, a specific rate is obtained by dividing R by the volume V of the system:

$$\boxed{r = \frac{1}{V}\frac{dX}{dt}} \tag{1.3.2}$$

Many sets of units are used to report reaction rates r in the literature. Frequently r is expressed in g-mole/liter-sec. Both definitions (1.3.1) and (1.3.2) are perfectly general and are valid for any region of space of uniform pressure, temperature and composition.

If, in addition, the volume of the system does not change with time, r will be designated as r_V. There are two expressions for r_V in common usage. Since in (1.3.2) V is now a constant, r can be rewritten as:

$$r_V = \frac{d\left(\dfrac{X}{V}\right)}{dt} = \frac{dx}{dt} \tag{1.3.3}$$

Thus the rate of reaction at constant volume is seen to be equal to the derivative of the conversion variable x with respect to time.

Also, differentiation of (1.2.2) gives

$$dn_i = \nu_i \, dX \tag{1.3.4}$$

If use is made of *molar concentrations* c_i of reactants or products:

$$c_i = \frac{n_i}{V} \tag{1.3.5}$$

substitution of (1.3.4) and (1.3.5) into (1.3.2) gives:

$$r_V = \frac{1}{\nu_i}\frac{dc_i}{dt} \tag{1.3.6}$$

Thus the rate of reaction at constant volume is seen to be proportional to the derivative of molar concentrations with respect to time. An unfortunate simplification which can lead to error is to set the proportionality coefficients, i.e., the stoichiometric coefficients, equal to unity. Then

$$r_V = -\frac{dc_i}{dt} \quad \text{for a specified reactant} \tag{1.3.7}$$

$$r_V = \frac{dc_i}{dt} \quad \text{for a specified product} \tag{1.3.8}$$

This practice should be avoided in general because the reactants or products for which these special definitions of the rate apply are easily forgotten. It must always be kept in mind that in any event, expressions for r_V should be used only when the volume of the system remains constant.

When it is not possible to write down a stoichiometric equation for the reaction, the simplified definitions (1.3.7) or (1.3.8) will suffice, or if the volume of the system does not stay constant, the corresponding expressions (1.3.9) and (1.3.10) will be used:

$$r = -\frac{1}{V}\frac{dn_i}{dt} \quad \text{for a specified reactant} \tag{1.3.9}$$

$$r = \frac{1}{V}\frac{dn_i}{dt} \quad \text{for a specified product} \tag{1.3.10}$$

For example, in addition polymerization such as styrene $\longrightarrow$ polystyrene for which no unique stoichiometric equation can be written, the rate is expressed as

$$r = -\frac{1}{V}\frac{dn}{dt}$$

where n is the number of moles of monomer. This must be compared to the general expression

$$(\nu_l)r = \frac{1}{V}\frac{dn_l}{dt} \tag{1.3.11}$$

derived from (1.3.2) and (1.3.4), the latter equation being applied to the limiting reactant l.

So far we have confined our attention to *homogeneous* reactions taking place in a closed system of uniform composition, temperature and pressure. But many reactions are *heterogeneous;* they take place at the interface between two fluid phases or between a fluid phase and a solid phase or between two solid phases. The general definition (1.3.1) of the extensive rate of reaction R remains valid. In order to obtain a convenient specific rate of reaction, it is now necessary to divide R by the interfacial surface area A available for reaction:

$$\boxed{r = \frac{1}{A}\frac{dX}{dt}} \tag{1.3.12}$$

The interfacial area entering in this definition must, of course, be of uniform composition, temperature and pressure. The rate r_A will then be expressed for instance in g-mole/cm^2-sec. In particular when the locus of reaction is an interface between a solid phase and a liquid phase, if the interfacial area, as frequently happens, is not known, alternative definitions of the specific rate are useful:

$$r = \frac{1}{W}\frac{dX}{dt} \tag{1.3.13}$$

$$r = \frac{1}{V'}\frac{dX}{dt} \tag{1.3.14}$$

where W and V' are weight viz volume of the solid particles dispersed in the fluid phase.

Of course many alternative definitions of specific rates for homogeneous and heterogeneous reactions are conceivable. Many of these are in use but conversion to the standard forms (1.3.2) and (1.3.12) should be attempted in every case where enough information is available. Whatever the choice of the definition of reaction rate adapted to a particular situation, it must be proportional to the derivative with respect to time of the extent of reaction.

As the extent of reaction is a thermodynamic variable, the general definition of the rate of reaction as dX/dt is also a thermodynamic quantity and indeed it plays a central role in the thermodynamics of irreversible processes. Being a thermodynamic quantity, it is totally unrelated to any molecular interpretation as to how the chemical reaction actually occurs. In particular, the definition applies to any *single reaction*, i.e., one the advancement of which can be described by a single variable. How this is possible in spite of the fact that the reaction normally takes place through reactive intermediates that do not appear in the equation for reaction, will be discussed in Chapter 3.

1.4 General Properties of the Rate Function for a Single Reaction

The rate of reaction is generally a function of temperature, composition and pressure. To find the form of the rate function r is a central problem of *applied chemical kinetics*. Once r is known, information can frequently be inferred on rates of individual steps for theoretical studies. Or a system, i.e., a reactor, can be designed for carrying out the reaction under optimum conditions.

In this paragraph, we present five general rules on the form of the rate function and much of the book will be devoted to the understanding and elucidation of these rules, of their range of validity, and of their exceptions. The rules are of an approximate nature but are sufficiently general that exceptions to them usually reveal something of interest. It must be stressed that the utility of these rules is their applicability to many single reactions. Application to elementary steps only would be far too restrictive and can be discussed readily in theoretical terms (see Chapter 2).

Rule I *The rate function r at constant temperature generally decreases in monotonic fashion with time or extent of reaction* (Fig. 1.4.1).

Rule II *The rate of an irreversible reaction can generally be written in the form:*

$$r = kF(c_i) \tag{1.4.1}$$

where $F(c_i)$ is a function that depends on the composition of the system as expressed by concentrations c_i.

The function $F(c_i)$ may also depend on temperature. The coefficient k does not depend on the composition of the system and therefore is independent of time. For this reason, k is called the *rate constant*.

Rule III *The rate constant generally depends on absolute temperature T following the law first proposed by Arrhenius* (1889):

$$\boxed{k = A \exp\left(-\frac{E}{RT}\right)} \tag{1.4.2}$$

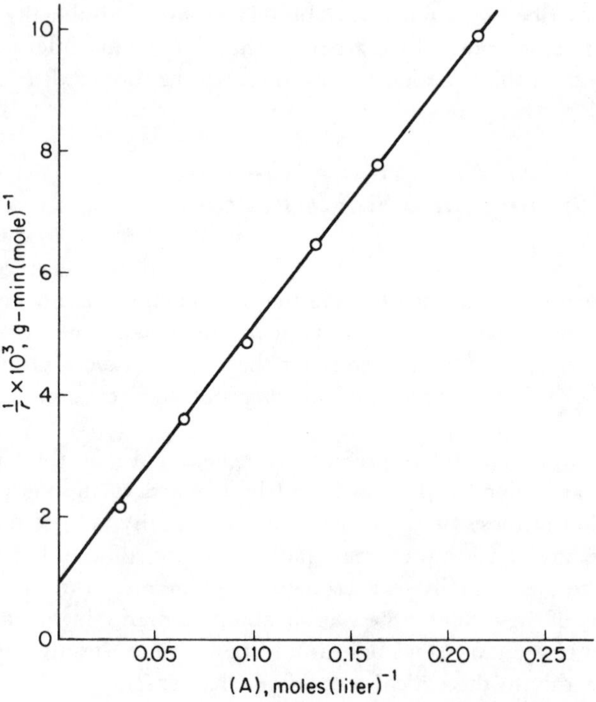

Fig. 1.4.1 The rate function r decreases with time. Liquid-phase dehydrogenation at 82°C, 1 atm, of isopropylalcohol with a nickel catalyst; (A) denotes concentration of acetone; the rate r is expressed in mole/minute·g catalyst. See D. E. Mears and M. Boudart, *AIChE Journal*, **12**, 313 (1966).

In this expression, the *pre-exponential factor* A does not depend on temperature. The *apparent activation energy* E is frequently expressed in g-cal/g-mole. Then the gas constant R has the value $R = 1.987 \cong 2$ g-cal/g-mole-degree C (Fig. 1.4.2).

Rule IV *Frequently, the function* $F(c_i)$ *in the expression* $r = kF(c_i)$ *is independent of temperature and to an excellent approximation can be written as:*

$$F(c_i) = \prod_i c_i^{\alpha i}$$

where the product Π is taken over all components of the system. The ex-

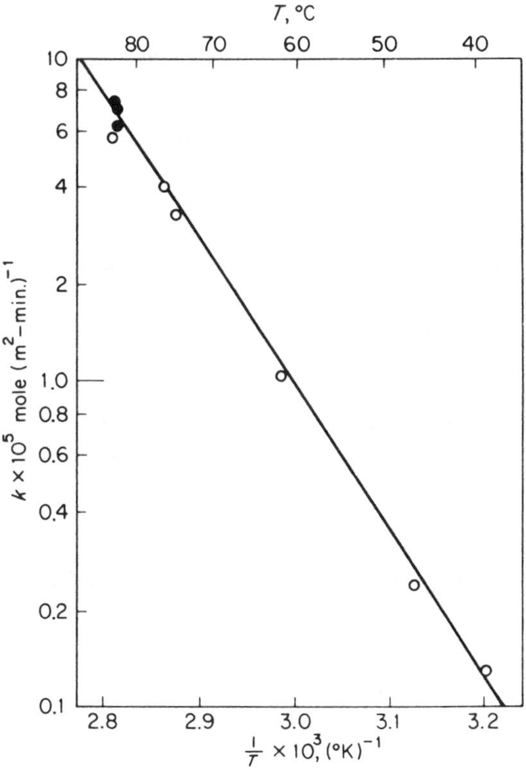

Fig. 1.4.2 A typical Arrhenius plot. For the dehydrogenation of isopropylalcohol, the rate r is expressed by:

$$r = \frac{k}{1 + \beta \text{(A)}}$$

where (A) is the concentration of acetone and β and k depend only on temperature (see Fig. 1.4.1). See D. E. Mears and M. Boudart, *AIChE Journal*, **12**, 313 (1966).

ponents α_i are small integers or fractions, positive, negative or zero. They are temperature independent, at least in a restricted interval.

Rule V *When a reaction is reversible, its rate can generally be expressed as the difference between a rate in the forward direction $\vec{r}$ and an opposing rate $\overleftarrow{r}$:*

$$r = \vec{r} - \overleftarrow{r} \tag{1.4.3}$$

When Rule II applies to the rate functions $\vec{r}$ and $\overleftarrow{r}$ so that $\vec{r} = \vec{k}\vec{F}(c_i)$ and $\overleftarrow{r} = \overleftarrow{k}\overleftarrow{F}(c_i)$, both rate constants $\vec{k}$ and $\overleftarrow{k}$ are related to the equilibrium constant K of the reaction by means of the equation:

$$\frac{\vec{k}}{\overleftarrow{k}} = K^n \tag{1.4.4}$$

The significance of the exponent n will be considered later.

The exponent α_i is said to be the *order of reaction* with respect to the corresponding component of the system. The algebraic sum of the exponents is called the *total order* of reaction.

It must be stressed that in general, α_i is not equal to the absolute value of the stoichiometric coefficient $|\nu_i|$ of the corresponding component in the equation for reaction. In fact α_i is rarely larger than two in absolute value. If α_i were equal to $|\nu_i|$ for reactants and equal to zero for all other components of the system, the expression

$$F(c_i) = \prod_i c_i^{|\nu_i|} \qquad \text{for reactants only}$$

would be of the form first suggested by Guldberg and Waage (1867) in their Law of Mass Action. In this book, we shall say that a rate function of the type:

$$\boxed{r = k \prod_i c_i^{\alpha_i}} \tag{1.4.5}$$

is in the *Guldberg-Waage form* rather than use other possible qualifications, e.g., "pseudo mass action" or "power law."

Some examples of Guldberg-Waage rate functions are collected in Table 1.4.1.

Problem 1.4.1

An ancient aphorism says: "The rate of reaction doubles for each ten degree Celsius rise in temperature." What is the corresponding value of the apparent activation energy for a reaction carried out in the vicinity of 300°K? Or in the vicinity of 500°K? By expanding k in Taylor series about any temperature T_0 and neglecting higher-order terms, find the factor by which k at T_0 is multiplied when T_0 is increased to a temperature in its vicinity.

Table 1.4.1

EXAMPLES OF RATE FUNCTIONS OF THE TYPE $r = k \prod_i c^{\alpha_i}$

Reaction	Rate Function
$CH_3CHO \rightarrow CH_4 + CO$	$k(CH_3CHO)^{1.5}$
$C_2H_6 + H_2 \rightarrow 2\ CH_4$	$k(C_2H_6)^{0.9}(H_2)^{-0.7}$
$SbH_3 \rightarrow Sb + \frac{3}{2}\ H_2$	$k(SbH_3)^{0.6}$
$N_2 + 3\ H_2 \rightarrow 2\ NH_3$	$k(N_2)(H_2)^{2.25}(NH_3)^{-1.5}$

1.5 *Ideal Reactors*

Systems in which chemical reactions take place are called reactors. Special names, such as reaction vessel, will not be used in this book. In practice, the situation in a reactor is usually very different from the ideal requirement used in the definition (1.3.1) of reaction rates. Indeed, a reactor is generally not a closed system with uniform temperature, pressure and composition. These ideal conditions will be rarely met even in experimental reactors designed for the measurement of reaction rates. In fact, reaction rates cannot be measured directly in a closed system; what *is* measured in a closed system is the composition of the system at various times and the rate is then inferred or calculated from these measurements.

There are several questions that can be raised about the operation of a reactor. They form the basis of a classification and define ideal conditions that are desirable for proper measurements of reaction rates. In practice, industrial reactors will approach these idealized limits more or less closely and the corresponding situations can be handled more or less successfully by the rapidly developing methods of *chemical engineering kinetics* (alias *chemical reaction engineering*).

The first question is whether the system exchanges mass with its surroundings. If it does not, we deal with a *static system* or *batch reactor*. If it does, the reactor is a *flow system* or *flow reactor*.

The second question concerns the exchange of heat between the system and its surroundings. If there is none, the reactor is *adiabatic*. At the other extreme, the reactor will make good thermal contact with a heat bath so that it is *isothermal*.

The behavior of the mechanical variables determines the third set of idealized operation: at constant *pressure* or at constant *volume*.

Fourth, the time spent in the reactor by each volume element of the reacting mass may be the same for all. This is the first simple alternative. On the other hand there may exist a *distribution of residence times* in which case, as will be seen in Section 1.6, the distribution turns out to be exponential for the opposite ideal situation to prevail.

Finally, if attention is focused on a fixed volume element in the reactor, all properties in that volume element may or may not change with time. If they do not, the reactor is said to operate at the *stationary state*. If they do, the system will operate under *transient* conditions. A nontrivial example of the latter situation obtains when a chemically reactive system at equilibrium is submitted to a perturbation and its kinetic behavior is deduced from the fact that the perturbed system returns or relaxes to equilibrium. The perturbation may be periodic. In any event, this is a typical *relaxation* experiment.

The ten possibilities that have just been outlined are collected in Table 1.5.1.

Table 1.5.1

LIMITING CONDITIONS OF REACTOR OPERATION

With respect to

exchange of mass	1. batch	2. flow
exchange of heat	3. isothermal	4. adiabatic
mechanical variables	5. constant volume	6. constant pressure
residence time	7. unique	8. exponential distribution
space-time behavior	9. transient	10. stationary

Problem 1.5.1

It would seem at first that a batch reactor must by necessity operate in transient fashion. In fact, it can be made to operate at the stationary state not in the strict sense but at least for all practical purposes. Try to design such an experimental reactor for the study of the gaseous reaction

$$N_2 + 3\,H_2 \rightarrow 2\,NH_3$$

taking place on a solid catalyst. Hint: Ammonia can be frozen out easily from a mixture of N_2, H_2 and NH_3 by passing the mixture through a cold trap.

1.6 Stirred-Flow Reactors

The ideal reactor for the direct measurement of reaction rates is a flow, isothermal, constant-pressure reactor operating at the stationary state with such thorough mixing that the composition is the same everywhere in the reactor. Because of its shape the reactor is frequently called a "stirred-tank reactor." If it operates at the stationary state it is sometimes called a *continuous flow stirred-tank reactor* (CFSTR) or more simply a *stirred-flow reactor*. In such a system, the composition in the reactor is ideally identical to that of the effluent stream and all the reaction therefore takes place at this constant composition of the effluent stream (Fig. 1.6.1).

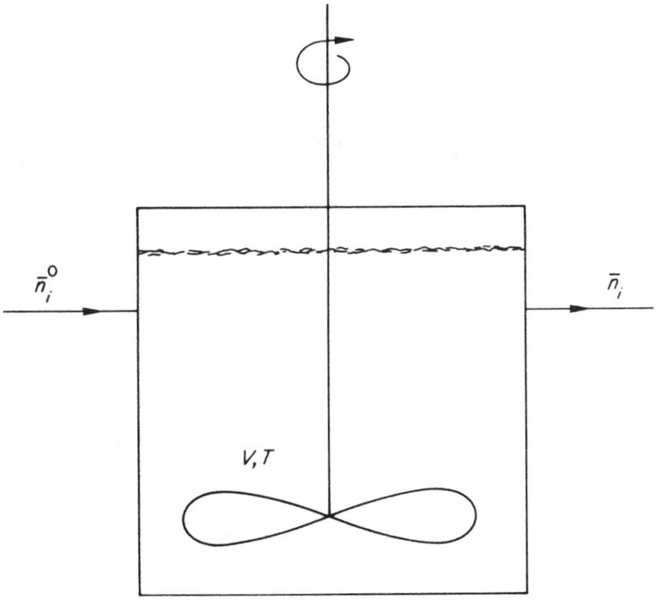

Fig. 1.6.1 Stirred-flow reactor. The composition of the reacting volume V at temperature T is the same everywhere and at all times.
$\bar{n}_i^0$: molar flow rate into the reactor
$\bar{n}_i$: molar flow rate out of the reactor

If then $\bar{n}_i^0$ denotes the number of moles of component i fed to the reactor per unit time and $\bar{n}_i$ is the number of moles of that component leaving the reactor per unit time, the difference between $\bar{n}_i^0$ and $\bar{n}_i$ must be due to the chemical reaction taking place at a rate r defined at the known constant temperature, and composition in the reacting system of volume V. With the help of Eqs. (1.3.2) and (1.3.4), the material balance can be written as:

$$\bar{n}_i - \bar{n}_i^0 = \nu_i V r \tag{1.6.1}$$

Therefore, the rate is measured directly:

$$\frac{\bar{n}_i - \bar{n}_i^0}{V} = \nu_i r \tag{1.6.2}$$

In particular, for the limiting reactant, we can write:

$$\frac{\bar{n}_l - \bar{n}_l^0}{V} = \nu_l r \tag{1.6.3}$$

On the other hand, according to (1.2.5):

$$f = \frac{n_l^0 - n_l}{n_l^0} = \frac{\bar{n}_l^0 - \bar{n}_l}{\bar{n}_l^0} \tag{1.6.4}$$

Substitution of (1.6.4) into (1.6.3) gives:

$$\boxed{\bar{n}_l^0 \frac{f}{V} = (-\nu_l) r} \tag{1.6.5}$$

If we write the equation for reaction in such a manner that $-\nu_l = 1$, we see that *the rate of reaction is equal to the number of moles of limiting reactant fed to the reactor per unit time and per unit volume of reactor, times the fractional conversion.*

For any product p not present in the feed stream we have:

$$\frac{\bar{n}_l^0}{-\nu_l} f = \frac{\bar{n}_p}{\nu_p}$$

where $\bar{n}_p$ is the number of moles of product p leaving the reactor per unit time. Thus Eq. (1.6.5) can be rewritten in the form:

$$\frac{\bar{n}_p}{V} = \nu_p r \tag{1.6.6}$$

The quantity $\bar{n}_p/V$ is called the *space-time yield*. The name is also used for other quantities proportional to $\bar{n}_p/V$.

If the equation for reaction is written in such a way that $\nu_p = 1$, it is seen that *in a stirred-flow reactor, the space-time yield of product p is simply equal to the rate of reaction.* This statement, which is unfortunately not true with all other types of chemical reactors, singles out stirred-flow reactors as the ideal tool of the kineticist for the measurement of rates of reaction.

The fundamental equations above describing the operation of stirred-flow reactors are valid whether the reaction takes place at constant volume or not. It is important to distinguish here carefully between V_r, the volume of the physical enclosure in which the system reacts and the volume V occupied by a given mass of the reacting system. Both are not necessarily equal. Furthermore, while it is clear that V_r is practically invariant, V almost always varies with extent of reaction in an isothermal system, except if the reaction mixture is an ideal gas contained in a batch reactor or passed through a flow reactor (provided that in the latter case the reaction is not accompanied by a change in number of moles).

In a stirred-flow reactor, the fluid reaction mixture is generally a liquid solution. If a large amount of inert solvent is used so that the solution is very dilute, the change in volume due to reaction can usually be neglected. But if the solution is concentrated or in the absence of solvent, V will depend on the extent of reaction. Clearly, the interpretation of data is simpler if V does not depend on the extent of reaction. It must be noted that V is not necessarily equal to V_r since the liquid phase will not normally fill up the entire reactor volume V_r available in principle for reaction.

If the change in volume due to reaction can be neglected, the operation of a stirred-flow reactor can be described in terms of the concentration c_l^0 of the limiting reactant l in the inlet stream, the concentration c_l of that reactant in the outlet stream and the volumetric flow rate $\overline{V}$ which must then be the same for inlet and outlet streams. Indeed a simple material balance gives for the number of moles of reactant l per unit time:

$$\overline{V}(c_l^0 - c_l) = V(-\nu_l)r_V \tag{1.6.7}$$

This can be rewritten as:

$$\boxed{\frac{c_l^0 - c_l}{\dfrac{V}{\overline{V}}} = (-\nu_l)r_V} \tag{1.6.8}$$

The ratio $V/\overline{V}$, i.e., the volume of mixture in the reactor divided by the volume of mixture fed to the reactor per unit time, is obviously a time with the simple interpretation of a nominal residence time. But the term *space-time* appears preferable because, in a stirred-flow reactor, there does not exist a single residence time but rather a distribution of residence times. The inverse of the space-time is called the *space velocity*. This latter term is also frequently used with other connotations, e.g., for catalytic reactors, it denotes the weight of feed per unit weight of catalyst per unit time. In every case, the conditions for the volume of feed must be specified: temperature, pressure (in the case of a gas) state of aggregation (liquid or gaseous). Space velocity or its inverse, the space-time, should be used in preference to the notions of "contact time" or "holding time." Indeed, as pointed out before, there is no unique residence time in the case of an ideal stirred-flow reactor.

This can be seen by performing one of a number of possible tracer experiments. Imagine a stirred-flow reactor at the stationary state through which flows a dilute water solution of a tracer substance. Its concentration c^0 is the same at inlet and outlet. At time $t = 0$, the feed is replaced by pure water at the same volumetric flow rate $\overline{V}$. How is the concentration of tracer in the reactor or in the effluent going to change with time? Again a material balance gives simply:

$$\overline{V}c = -V\frac{dc}{dt} \tag{1.6.9}$$

Integration with boundary condition $c = c_0$ at $t = 0$ yields:

$$c = c_0\left[e^{-\frac{t}{(V/\overline{V})}}\right] \tag{1.6.10}$$

This exponential decay is typical of all first-order processes (e.g., radioactive decay): the rate constant k of such processes:

$$-\frac{dc}{dt} = kc \tag{1.6.11}$$

is replaced in (1.6.10) by the space velocity $\overline{V}/V$.

A first-order rate law of this type has a simple meaning. Consider n particles in a system. If the probability of disappearance of each one is independent of the presence of the others and equal to k per unit time, the number of particles in the system decays at the rate:

$$-\frac{dn}{dt} = kn \tag{1.6.12}$$

In integrated form and in terms of the fractional conversion f, we get

$$\ln\frac{1}{1-f} = kt \tag{1.6.13}$$

The time required for the disappearance of half of the particles is called their half-life $t_{1/2}$. It is given by:

$$t_{1/2} = \frac{\ln 2}{k} = \frac{0.69}{k} \tag{1.6.14}$$

Therefore the half-life of a particle in a stirred-flow reactor is equal to 0.69 times the space-time. Some molecules will spend very little time in the reactor; some will stay very long. But the mean residence time $<t>$ is equal to the nominal residence time $V/\bar{V}$. Indeed:

$$<t> = \frac{\int_0^\infty tc(t)dt}{\int_0^\infty c(t)dt} = \frac{V}{\bar{V}} \tag{1.6.15}$$

This result is obtained by substituting under both integral signs $c(t)$ by its value as given by (1.6.10) and noting that $\int_0^\infty xe^{-x} = 1$.

By applying the same reasoning to a first-order process (1.6.11), we see that the result equivalent to (1.6.15) says that the mean lifetime of a particle in a collection of particles that decay following a first-order rate law is simply the inverse of the rate constant k. Similarly, as found in (1.6.14) the half-life of a particle is its mean life multiplied by 0.69. Thus any first-order rate constant receives a very simple physical meaning.

Problem 1.6.1

The space-time yield is a measure of utilization of reactor volume. If Rule I concerning a rate function r obtains, explain for the simple case of a stirred-flow reactor why there would exist an economic balance between the desirability of obtaining both high conversions and a large space-time yield.

Problem 1.6.2

The rate of the reaction:

$$OH^- + CH_3COOC_2H_5 \rightarrow CH_3COO^- + C_2H_5OH$$

has been found to be first-order with respect to hydroxyl ions and first-order with respect to ethyl acetate. In a stirred-flow reactor of volume $V = 602$ ml, the following data have been obtained at 25°C by Denbigh, et al. [*Disc. Faraday Soc.*, **2**, 263 (1947)]:

Flow rates at inlet: solution of barium hydroxide: 1.16 liter/hr.
solution of ethyl acetate: 1.20 liter/hr.
Concentrations: (OH^-) inlet: 0.00587 mole/liter.
($CH_3COOC_2H_5$) inlet: 0.0389 mole/liter.
(OH^-) outlet: 0.001094 mole/liter.

Calculate the rate constant. Changes of volume accompanying the reaction are negligible.

Problem 1.6.3

Consider a rate function of the form:

$$-\frac{dc}{dt} = kc^2$$

What is the relation between the second-order rate constant k and the half-life $t_{1/2}$?

Problem 1.6.4

Stirring gases is inconvenient and inefficient at any scale. For the study of gas phase reactions catalyzed at the surface of a solid, the mechanical problem of setting up a gas-solid stirred-flow reactor is a difficult one. If, however, a recirculation pump for gases is available, a stirred-flow reactor can be built readily for measuring the rate of a reaction catalyzed by a solid. Show how.

1.7 *Ideal Tubular Reactors*

Another simple type of flow reactor is the tubular reactor operating isothermally, at constant pressure, at the stationary state and with a unique residence time. This reactor is a cylindrical pipe of constant cross-section completely filled with reaction mixture which flows as if it were a rigid plug or as if it were pushed by a piston. Hence the name of plug or piston flow reactor is given to it. In this reactor, composition is the same at all points in a given cross-section but changes along the axial coordinate L of the pipe. If it is assumed that no mixing takes place between adjacent volume elements either radially or axially, either by diffusion or convection, all volume elements entering the reactor clearly have the same residence time. (Fig. 1.7.1)

Consider a section of reactor of length dL. Material balance per unit time for the limiting reactant in the volume element $S(dL)$ where S is the cross-sectional area gives:

$$d\bar{n}_l = \nu_l S(dL) r \qquad (1.7.1)$$

But (1.6.4) can be rewritten as:

$$\bar{n}_l = \bar{n}_l^0(1 - f)$$

where as before $\bar{n}_l^0$ is the number of moles of l entering the reactor per unit time. We get then:

$$\bar{n}_l^0 \frac{df}{S(dL)} = (-\nu_l)r \qquad (1.7.2)$$

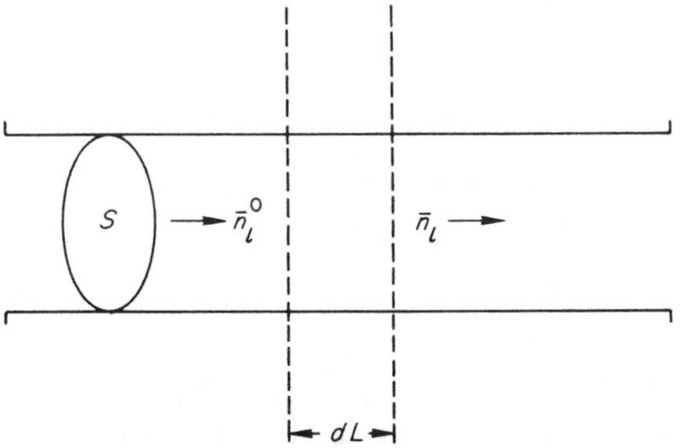

Fig. 1.7.1 Tubular reactor.

But $S(dL) = dV_r$ where V_r is reactor volume. Thus:

$$\boxed{\bar{n}_l^0 \frac{df}{dV_r} = (-\nu_l)r} \qquad (1.7.3)$$

If changes in volume due to reaction are negligible, the same reasoning that led to (1.6.8) gives:

$$\boxed{\frac{dc_l}{d\left(\frac{V_r}{\bar{V}}\right)} = \nu_l r} \qquad (1.7.4)$$

This last equation is identical to that of a batch, isothermal, constant-volume, transient batch reactor with a unique residence time t:

$$\boxed{\frac{dc_l}{dt} = \nu_l r} \qquad (1.7.5)$$

Clearly, space-time $V_r/\bar{V}$ in the ideal tubular reactor is the same as residence time in the ideal batch reactor but this is true, even in the case of ideal operation, only if the reaction is not accompanied by a volume change. Otherwise, the space-time calculated with volumetric flow rate at entrance is not equal to residence time. The latter depends on degree of conversion and is therefore not a useful concept. In general, with flow reactors, residence times should be used with caution since there may be a distribution of them or they may depend on conversion.

To fix the ideas and illustrate the difference between V_r and V, consider a *pure gas* fed to an ideal tubular reactor operating at temperature T and pressure p. The stoichiometric equation is:

$$0 = \nu_l \mathrm{A}_l + \Sigma\, \nu_p \mathrm{A}_p \tag{1.7.6}$$

The sum of the stoichiometric coefficients is $\nu_l + \Sigma\, \nu_p = \nu$. As a result of reaction the number of moles changes from n_l^0 to $n_l^0[1 + (\nu/-\nu_l)f]$. The rate function is

$$r = kc_l \tag{1.7.7}$$

Unless ν is equal to zero, the number of moles changes with extent of reaction and since p and T are constant, the volume V available to the reacting species changes with extent of reaction according to

$$V = n_l^0\left(1 + \frac{\nu}{-\nu_l}f\right)\frac{RT}{p} \tag{1.7.8}$$

if we deal with a mixture of ideal gases. The concentration

$$c_l = \frac{n_l}{V} \tag{1.7.9}$$

changes therefore on two counts: because of a change in number of moles n_l and also because of a change in volume V.

Using (1.7.7), (1.7.8), and (1.7.9) and substituting into (1.7.3) we get:

$$\bar{n}_l^0 \frac{df}{dV_r} = (-\nu_l)\frac{kp}{RT}\frac{(1-f)}{\left(1 + \dfrac{\nu}{-\nu_l}f\right)} \tag{1.7.10}$$

The volume of reactor V_r required to reach a degree of conversion f for a molar flow rate $\bar{n}_l^0$ is by integration of (1.7.10):

$$V_r = \frac{\bar{n}_l^0 RT}{(-\nu_l)kp}\left[\int_0^f \frac{df}{1-f} + \int_0^f \frac{\dfrac{\nu}{-\nu_l}f}{(1-f)}\,df\right] \tag{1.7.11}$$

The first term in the bracket would be the only one there if ν were equal to zero, i.e., if the reaction took place at constant volume. If however ν is positive, i.e., if the reaction is accompanied by an increase of volume, the volume of reactor will have to be larger because of the contribution of the second term. This can be understood readily since as the gas moves through the reactor, its linear velocity will increase. Therefore the residence time will be smaller than would be the case if no volume change took place.

Problem 1.7.1

Derive an equation similar to (1.7.2) for the case of an ideal tubular reactor packed with a solid catalyst. The surface area of the catalyst has the value A per unit length of reactor.

Problem 1.7.2

An equimolar mixture of ammonia and nitrogen is fed to an ideal isothermal tubular reactor packed with 5 grams of a catalyst with a specific surface area equal to 200 m^2/gram.

The total feed rate to the reactor is 3.6 liter STP/hr. The catalyst decomposes the ammonia and the degree of conversion of the latter is observed to be 0.5. If it is assumed that the rate of decomposition of ammonia is first-order with respect to ammonia, what is the value of the rate constant?

In the first part of your answer, explain how to attack such a problem quite generally.

In the second part, explain how to apply the general solution to the case at hand.

In the third part, present the numerical solution to the problem.

1.8 Measurement of Reaction Rates

The main relations derived for ideal reactors are collected in Table 1.8.1. For measurements of reaction rates, a deliberate effort will be made to operate under the set of ideal conditions of these reactors. While, in practice, large-sized reactors will exhibit more or less pronounced deviations from these *ideal* conditions, it is often interesting to calculate the volume of the ideal reactor required for a given yield by use of the equations of Table 1.8.1. This can be done if the rate functions r or r_V are known.

Calculation of the volume of reactor required under *real* conditions is a central problem of chemical reaction engineering. But in these applications the rate function r remains the most important information required. This book will be devoted largely to the rate function r, its form and its meaning. Alternatively, the formulae collected in Table 1.8.1 are those that may be used to obtain the rate function from kinetic measurements in ideal reactors.

Table 1.8.1

RATE OF HOMOGENEOUS REACTIONS IN IDEAL ISOTHERMAL REACTORS

Stirred-Flow	Ideal Tubular	Batch
$\bar{n}_l^0 \frac{f}{V} = (-\nu_l) r$	$\bar{n}_l^0 \frac{df}{dV_r} = (-\nu_l) r$	$\frac{1}{V} \frac{dn_l}{dt} = \nu_l r$
(1.6.5)	(1.7.3)	(1.3.11)
$\dfrac{c_l - c_l^0}{\dfrac{V}{\bar{V}}} = \nu_l r_V$	$\dfrac{dc_l}{d\left(\dfrac{V_r}{\bar{V}}\right)} = \nu_l r_V$	$\frac{dc_l}{dt} = \nu_l r_V$
(1.6.8)	(1.7.4)	(1.7.5)

Nomenclature

c_l: molar concentration of limiting reactant: g-mole/liter
c_l^0: initial molar concentration of limiting reactant: g-mole/liter
f: fractional conversion: dimensionless
n_l: number of moles of limiting reactant: g-mole
$\bar{n}_l^0$: molar feed rate of limiting reactant: g-mole/sec
r: reaction rate: g-mole/liter-sec
r_V: reaction rate at constant volume: g-mole/liter-sec
ν_l: stoichiometric coefficient of limiting reactant: dimensionless
V: volume of reacting system: liter
$\bar{V}$: volumetric flow rate: liter/sec
V_r: volume of reactor: liter

Problem 1.8.1

A first-order reaction takes place isothermally and at constant volume in two reactors of identical volume. One is a stirred-flow reactor and the other is an ideal tubular reactor. If the space-time is the same in both cases, in which reactor will the space-time yield be higher and why?

*BIBLIOGRAPHY**

1.2 The extent of reaction is a parameter first introduced by De Donder. It has been used systematically by the Brussels School of Thermodynamics. See I. Progogine and R. Defay (tr. D. H. Everett), *Chemical Thermodynamics*, Longmans, Green and Co., London, 1954.

***Boldface numbers to the left of bibliographical entries refer to the pertinent section of the preceding chapter.**

1.3 The use of volume (usually a variable) in the definition of reaction rate is an unfortunate tradition in chemical kinetics. So is the use of concentrations (moles per unit volume). Alternative, more desirable schemes have not gained favor so far. See S. S. Penner, *Chemical Reactions in Flow Systems*, Butterworths, London, 1955. For a recent discussion of r vs r_V, see Max S. Peters and Edward J. Skorpinski, *J. Chem. Ed.*, **42**, 329 (1965).

1.5 A broad survey, with valuable historical references, of "Progress towards the *a priori* design of chemical reactors" is found in a lecture by R. H. Wilhelm, *Pure and App. Chem.*, **5**, 403 (1962).

1.6 The concepts of space-time yield and space velocity are ancient. They are discussed in an early book: H. S. Taylor and E. K. Rideal, *Catalysis in Theory and Practice*, London, 1919. Space-time has been introduced by O. Levenspiel who treats in detail the principles sketched here in his book *Chemical Reaction Engineering*, John Wiley & Sons, Inc., New York, 1962. The history of stirred-flow reactors is well covered, with many examples by William C. Herndon, *J. Chem. Ed.*, **41**, 425 (1964).

1.7 Many of the concepts sketched in this section and the two preceding ones are discussed very lucidly by Rutherford Aris in his text *Introduction to the Analysis of Chemical Reactors*, Prentice-Hall, Inc., Englewood Cliffs, N.J., 1965.

1.8 "Fundamental Operations and Measurements in Obtaining Rate Data," "Time Measurements and the Recording of Kinetic Data," and "Evaluation and Interpretation of Rate Data," are discussed in *Technique of Organic Chemistry*, Vol. VIII, Part I, S. L. Friess, E. S. Lewis and A. Weissberger, eds., Interscience Publishers, New York, 1961.

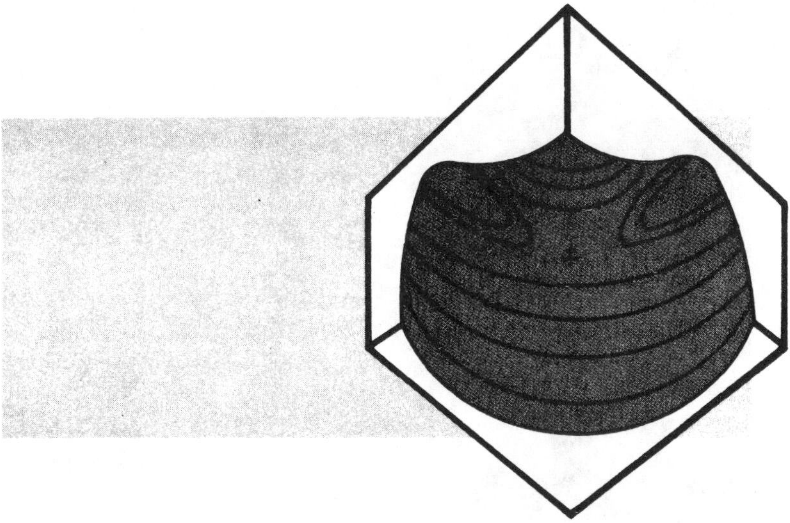

Chemical Kinetics of Elementary Steps

2

2.1 *Definition of an Elementary Step*

A reaction is an elementary step if it takes place in a single *irreducible* act at the molecular level, just the way it is written in the stoichiometric equation for reaction. The act of reaction is irreducible when no intermediate between reactants and products can be detected by any available technique. Clearly, the word "irreducible" has an operational meaning: What appeared at one time to be elementary, today may well be broken down into two or more simpler steps.

Indeed, simplicity has been elevated to the rank of principle, that of the least change of structure (Muller, 1886). The act of reaction will normally be a simple one, where only one bond is made or broken. Also simple is the situation where one bond is broken while another one is made. More rarely, two bonds are broken and two new ones are made in what is commonly called a four-center reaction. While more complicated rearrangements are known to occur in one step consisting of a series of concerted motions, they must be looked upon with suspicion in the absence of independent evidence which shows conclusively that they are truly irreducible. Thus the following reaction:

$$OH + CH_4 \rightarrow CH_3 + H_2O$$

is a good candidate for an elementary step, while the reaction

$$CH_3 + O_2 \rightarrow OH + CH_2O$$

can hardly qualify without further examination since it requires breaking two bonds and making two bonds. In the final analysis, whether a reaction is an elementary step or not must be decided by experiment. As an illustration, Table 2.1.1 presents Burwell's classification of elementary steps taking place

Table 2.1.1

Classification of Elementary Steps at Catalytic Surfaces Following Burwell

Step	Bonds Broken	Bonds Made
I. Adsorption and its reverse, desorption		
(a) $* + NH_3(g) \rightarrow H_3^+N*^-$	none	A—*
(b) $* + H(g) \rightarrow H*$	none	A—*
II. Dissociative adsorption and its reverse, associative desorption		
(a) $2* + H_2(g) \rightarrow 2H*$	A—A	2(A—*)
(b) $2* + CH_4(g) \rightarrow CH_3* + H*$	A—B	A—*, B—*
(c) $2* + CH_2{=}CH_2(g) \rightarrow *CH_2CH_2*$	A—A	2(A—*)
III. Dissociative surface reaction and its reverse, associative surface reaction		
(a) $2* + C_2H_5* \rightarrow H* + *CH_2CH_2*$	A—B*	A—*, B—*
(b) $2* + *CH_2CH_2CH_2CH_2* \rightarrow 2*CH_2CH_2*$	*A—A*	2(A—*)

Table 2.1.1—*Cont.*

Step	Bonds Broken	Bonds Made
IV. Reactive adsorption and its reverse, reactive desorption		
(a) $H* + C_2H_4(g) \rightarrow C_2H_5*$	A—*, B—B	A—B*, B—*
(b) $H_2C{=}CH_2 + D{-}D(g) \rightarrow$ 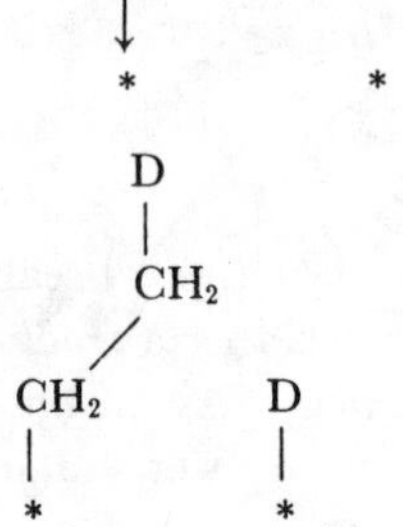	?, A—A	A—B*, B—*, A—*
(c) $*CH_2CH_2* + H_2(g) + * \rightarrow *C_2H_5 + * + H*$	B—*, A—A	A—B*, A—*
V. Quasisorbed reaction		
(a) $H(g) + H* \rightarrow H_2(g) + *$	A—*	A—A
(b) $2H* + C_2H_4(g) \rightarrow 2* + C_2H_6(g)$	2(A—*), B—B	2(A—B)
(c) $H* + D_2(g) + * \rightarrow * + HD(g) + D*$	A—*, B—B	B—*, A—B
(d) $D* + H_2C{=}CH{-}CH_3(g) \rightarrow DH_2C{-}CH{=}CH_2(g) + H*$	A—*, B—C D—E	E—*, A—B C—D

at a catalytic surface presenting active sites denoted by an asterisk. The list does not include all possibilities and it is not unlikely that some of the reactions in the table, e.g., IIa, will be shown not to be elementary as a result of further work.

Since an elementary step represents a molecular event, its equation for reaction may not be written arbitrarily. Rather it must be written the way it takes place. Many errors are perpetrated for lack of adherence to this rule. Thus again, Example IIa in the table, the adsorption of molecular hydrogen at the surface of a solid catalyst to give two adsorbed hydrogen atoms, may not be written as:

$$* + \tfrac{1}{2} H_2 \rightarrow H*$$

since there is no such thing as half a molecule of hydrogen.

Clearly most reactions of interest cannot be elementary steps and it would seem that a general theory that encompasses only the latter would have only limited practical significance. Such is not the case because a reaction generally takes place in a sequence of elementary steps, and its rate can be predicted in principle from a knowledge of the rates of its component steps.

The study of elementary steps is the proper domain of *pure chemical kinetics*. The general theory of the rate of elementary steps is absolute rate theory or the *theory of the transition state*. As elementary steps can be so varied from a chemical standpoint it might seem that a general theory of their rates would be, if it exists at all, so general as to be almost useless. Such is not the case. The main results of this general theory will be presented insofar as they provide an answer to the basic question: How does the rate of an elementary step depend on temperature, pressure (or volume) and composition of the system?

2.2 Transition-State Theory

For an extremely large class of elementary steps, the rearrangement of atoms occurs through the motion of nuclei in the continuous potential field set up by the rapid motion of the electrons of the system. In other words, for the elementary step:

$$A + B \rightleftarrows C + D$$

there exists in multidimensional space a single potential energy surface on which the system will move to go from reactants to products and back. On this potential energy surface there exists a path between reactants and products that will be the most economical in terms of the energy required for reaction. Position along the reaction path is determined by the value of a *reaction coordinate*. Then, if the potential energy of the system is plotted versus the reaction coordinate, a schematic two-dimensional diagram of the type shown in Fig. 2.2.1 must be obtained.

The calculation of a potential energy surface for reaction or of a diagram such as that of Fig. 2.2.1 is a formidable, yet unsolved problem in quantum mechanics. After more than thirty years of efforts, the solution of this problem, obtained with the help of digital computers, now appears in sight for the simplest conceivable reaction: $H + H_2 \rightarrow H_2 + H$. A recent potential energy surface for this elementary step is shown in Fig. 2.2.2. But a mere knowledge of the general topography of the system is all that is necessary to derive a number of general conclusions. It is clear that regions of low energy corresponding to products and reactants are as a rule separated by a region of high energy. The highest energy along the most economical reaction path defines the *transition state*, also called *activated complex*. It must be stressed that

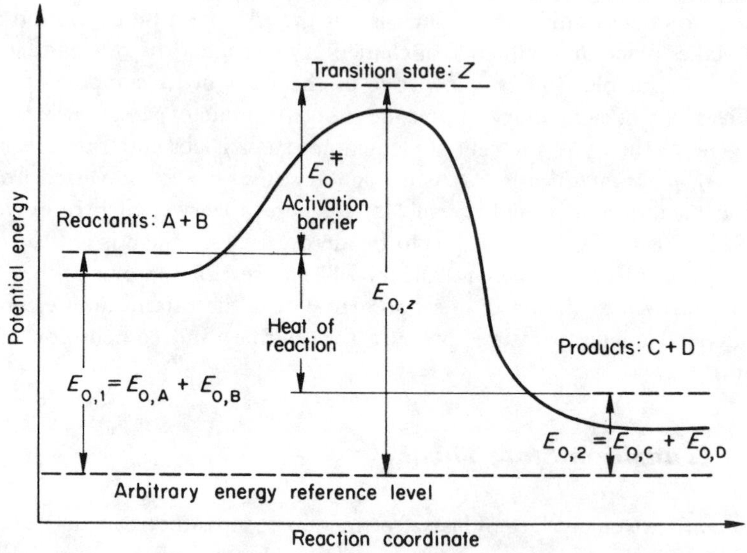

Fig. 2.2.1 Potential energy profile for elementary step:

$$A + B \rightarrow C + D$$

The reaction $A + B \rightarrow C + D$ is exothermic from left to right. Its heat of reaction at 0°K is: $E_{0,1} - E_{0,2}$. The activation barrier is: $E_{0,Z} - E_{0,A} - E_{0,B}$. For the reverse endothermic reaction, $C + D \rightarrow A + B$, the activation energy is the sum of the heat of reaction and activation barrier.

the transition state is not an intermediate; it corresponds merely to a special configuration of a system in transit from one state to another.

The difference in energies between reactants and products is the heat of reaction, a purely thermodynamic quantity. The elementary step pictured on Fig. 2.2.1 is exothermic from left to right and endothermic from right to left. For an exothermic step, the difference in energies between transition state and reactants is a purely kinetic quantity called the *activation barrier*. For an exothermic step, the activation barrier is identical to the *activation energy*. For an endothermic step, the activation energy is the sum of the activation barrier and the heat of reaction.

According to quantum mechanics, a molecular system at 0°K will possess a residual energy called the *zero-point energy*. In Fig. 2.2.1, zero-point energies of reactants, products and transition state have been represented, as well as activation barrier, activation energy and heat of reaction, all at absolute zero.

The task of transition-state theory is to calculate the rate of an elementary step that conforms to the general situation depicted on Fig. 2.2.1.

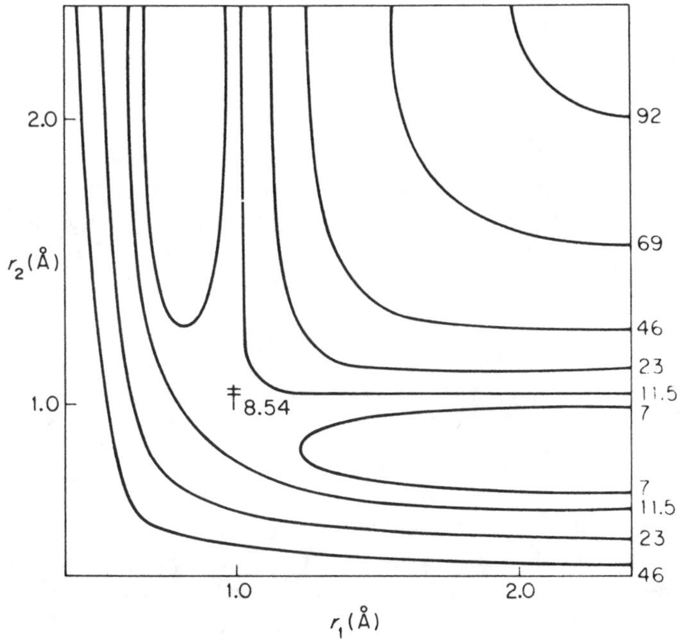

Fig. 2.2.2 The potential energy surface for the reaction:

$$H + H_2 \longrightarrow H_2 + H$$

The lines of equal energy correspond to energies in kcal/g-mole. r_1 is the distance between approaching H and the nearest H of reactant H_2 in collinear approach:

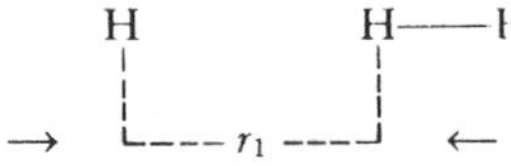

r_2 is similarly the distance between receding H and the nearest H of product H_2 in collinear motion:

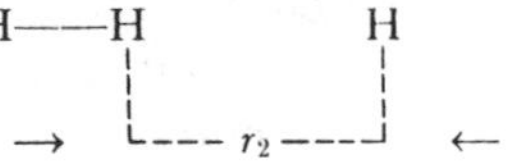

[from R. N. Porter and M. Karplus, *J. Chem. Phys.*, 40, 2358 (1964)]

2.3 *The Rate of an Elementary Step*

Consider the step $A + B \leftrightarrow C + D$ in a system at thermodynamic equilibrium. The rate of reaction from left to right $\vec{r}$ must then be equal to the rate of reaction from right to left $\overleftarrow{r}$. We shall use the symbol $\leftrightarrow$ to denote a reaction at equilibrium, i.e., a reaction for which $\overrightarrow{r} = \overleftarrow{r}$. By contrast the symbol $\rightleftarrows$ will be used for reversible reactions not at equilibrium while a simple arrow, $\rightarrow$, will denote an irreversible reaction, i.e., one for which $\overleftarrow{r} \ll \overrightarrow{r}$.

In our equilibrated system A + B $\leftrightarrow$ C + D, there exists a certain equilibrium concentration c_Z of transition states. In this entire chapter, concentration in mixtures of perfect gases will be expressed as a number density, i.e., number per unit volume. The rate of reaction is equal to the number of systems per unit volume crossing the barrier per unit time:

$$\overrightarrow{r} = \overleftarrow{r} = \nu c_Z \tag{2.3.1}$$

where ν is a frequency. An observer at the top of the barrier would determine this frequency ν at which transition states cross the barrier, going from reactants to products or vice versa.

Since c_Z is a thermodynamic quantity, its calculation can be made, in principle, by the methods of statistical thermodynamics. This is an enormous simplification of the kinetic problem. The fundamental assumption of transition-state theory is that if now the products are removed from the system at equilibrium, the rate of the reaction in one direction, A + B $\rightarrow$ C + D, is still given by the expression (2.3.1) prevailing at equilibrium:

$$\boxed{r = \overrightarrow{r} = \nu c_Z} \tag{2.3.2}$$

where c_Z is the concentration of transition states in equilibrium with reactants A and B. In more advanced treatments, it is shown that the *fundamental assumption merely requires that reactants be in thermodynamic equilibrium among themselves*. The assumption of equilibrium is believed to be generally valid. Its meaning and its limitation are best illustrated by a related problem which played an important role in the early days of the kinetic theory of gases.

This is the problem of the rate of evaporation of a liquid into a vacuum, first considered by Hertz (1882). Consider a liquid, such as mercury, in equilibrium with its vapor in a container kept at an absolute temperature T. At equilibrium (Fig. 2.3.1a) the number of vapor molecules striking unit liquid surface per unit time and condensing upon it (i.e., the rate of condensation r_c) is equal to the number of molecules leaving the liquid-vapor interface, per unit time and per unit area (i.e., the rate of evaporation r_e). If a molecule striking the liquid from the vapor condenses on the surface with a probability α, the rate of condensation r_c can be related to the rate of collision of vapor molecules with the surface r_{coll} through this condensation coefficient α:

$$r_c = \alpha r_{\text{coll}} \tag{2.3.3}$$

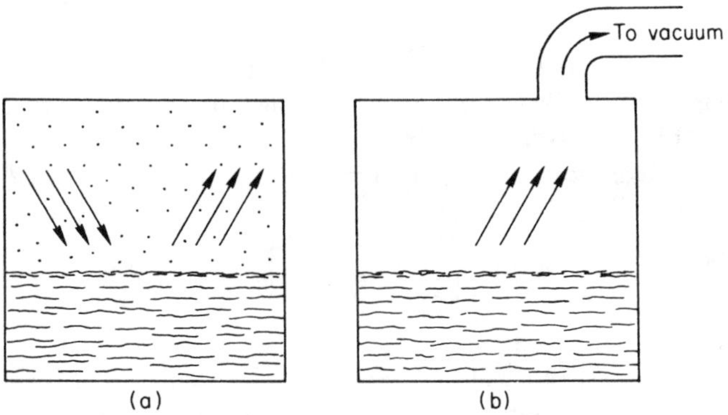

The liquid is in equilibrium with its vapor.

The liquid evaporates into a vacuum.

Fig 2.3.1 Evaporation of a liquid.

In particular the maximum rate of condensation $r_{c,\max}$ will correspond to a probability of condensation equal to unity so that

$$r_{c,\max} = r_{\text{coll}} \tag{2.3.4}$$

But the rate of collision of gas molecules with a surface is a familiar result of the kinetic theory of gases; the number of molecules striking a unit of surface area per unit time is proportional to the mean molecular velocity v and to the number density c of molecules in the gas:

$$r_{\text{coll}} = \tfrac{1}{4}vc \tag{2.3.5}$$

Again from kinetic theory of gases, it is recalled that the mean velocity of molecules in a gas is almost equal to the velocity of sound in the gas and is given by:

$$v = \left(\frac{8kT}{\pi m}\right)^{1/2} \tag{2.3.6}$$

where m is the mass of the molecule and k is Boltzmann's constant.

Still at equilibrium, to a maximum rate of condensation will correspond a maximum rate of evaporation, $r_{e,\max}$:

$$r_{e,\max} = r_{c,\max} = r_{\text{coll}} = (\tfrac{1}{4})vc_e \tag{2.3.7}$$

where c_e denotes the number density of vapor molecules in equilibrium with the vapor.

Suppose now that the vapor space above the liquid is connected to an efficient vacuum pump so that the molecules leaving the liquid surface never return to it (Fig. 2.3.1b). According to the fundamental assumption of equilibrium, the maximum rate of evaporation will still be given by (2.3.7):

$$\boxed{r_{e,\max} = \tfrac{1}{4}vc_e} \tag{2.3.8}$$

In other words, the logical step between (2.3.1) and (2.3.2) is very similar to that between (2.3.7) and (2.3.8). It will be noted that the rate of evaporation, according to (2.3.8), is still expressed in terms of an equilibrium concentration c_e which has been made nonexistent by the pumping away of the gas molecules. Similarly, the rate of reaction r is expressed in terms of an equilibrium concentration c_Z of transition states, following (2.3.2) even though they too are being pumped away to reaction products in irreversible fashion.

It is seen that the problem of Hertz is of more than historical interest. In the case of mercury, the acceptance of the equilibrium assumption has been fully justified by the definitive work of Volmer and Estermann (1922) who showed that the coefficient α was equal to unity or that the actual rate of evaporation was equal within experimental error to the maximum rate given by (2.3.8).

Another useful parallel between the rate of evaporation and the rate of reaction is that concerning a possible limitation to the assumption of equilibrium. When a liquid evaporates into a vacuum, the rate of evaporation can be so high that the surface layers are severely cooled because of an insufficient supply of heat from the bulk of the liquid in the bath at constant temperature. Similarly, in the case of an extremely rapid reaction, the most energetic molecules among reactants are depleted by reaction and a situation may develop in which energy transfer among reactants is not fast enough to maintain an equilibrium distribution of energy among reactants. In both cases, the assumption of equilibrium would fail. The success of the assumption of equilibrium in the case of reaction rates can thus be ascribed to the fact that normally rates of energy transfer in reacting systems are more than adequate to keep up with rates of reaction.

Problem 2.3.1

Calculate the rate of evaporation of liquid mercury into a vacuum. The temperature is 25°C. Find the necessary information in any handbook.

2.4 Thermodynamic Formulation of Rates

As a consequence of the fundamental assumption of equilibrium, everything happens as if transition states Z at concentration c_Z were in equilibrium with reactants A and B for an elementary step A + B → Products.

To the fictitious equilibrium A + B ↔ Z corresponds an equilibrium constant:

$$K_c^{\ddagger} = \frac{c_Z}{c_A c_B} \tag{2.4.1}$$

with the usual thermodynamic relation:

$$RT \ln K_c^{\ddagger} = -\Delta G^{0\ddagger} = -\Delta H^{0\ddagger} + T\Delta S^{0\ddagger} \tag{2.4.2}$$

where R is the gas constant, $\Delta G^{0\ddagger}$ is the change in standard Gibbs free energy for the reaction A + B → Z while $\Delta H^{0\ddagger}$ and $\Delta S^{0\ddagger}$ are the corresponding changes in standard enthalpy and standard entropy. The superscript 0 is used to denote standard states while superscript ‡ denotes quantities pertaining to the formation of the transition state considered as a molecular species which is assumed to possess normal thermodynamic properties. The transition state can be treated as a normal molecular species except that one of its vibrational modes is missing and must be replaced by translation along the reaction coordinate.

Now with the help of (2.4.1) and (2.4.2), (2.3.1) becomes

$$r = \nu \exp\left(\frac{\Delta S^{0\ddagger}}{R}\right) \exp\left(\frac{-\Delta H^{0\ddagger}}{RT}\right) c_A c_B \tag{2.4.3}$$

From the preceding section, it must be clear that the fundamental assumption of equilibrium is not characteristic of transition-state theory. The same assumption has been used in other rate theories, in particular, theories based on the kinetic theory of gases.

Rather the fundamental assumption of transition-state theory must be considered to be the following:

The frequency ν in Eq. (2.4.3) is a universal frequency, i.e., it does not depend on the nature of the molecular system considered.

In order then to determine ν once and for all, it is sufficient to identify this frequency for only one case. This will be done presently and it will be shown that

$$\boxed{\nu = \frac{kT}{h}} \tag{2.4.4}$$

where h is Planck's constant.

The identification of ν can be carried out readily for a hypothetical reaction consisting of the collision between two hard spheres A and B. Indeed the rate of collision between two such molecular species of diameters σ_A and σ_B and of masses m_A and m_B is well known from gas kinetic theory:

$$\boxed{r_{\text{coll}} = \pi\sigma^2 v_\mu c_A c_B} \tag{2.4.5}$$

where σ is the mean molecular diameter $(\sigma_A + \sigma_B)/2$ and v_μ is the mean molecular velocity

$$v_\mu = \left(\frac{8kT}{\pi\mu}\right)^{1/2} \tag{2.4.6}$$

calculated by means of the reduced mass

$$\mu = \frac{m_A m_B}{m_A + m_B} \tag{2.4.7}$$

We will now calculate r_{coll} by means of (.24.3) and verify (2.4.4) by comparing the result with (2.4.5).

In order to use (2.4.3), we shall need expressions from statistical mechanics for the entropy of translation S_t for the entropy of rotation S_r of a rigid diatomic rotator.

The entropy of translation of an ideal gas is given by the Sackur-Tetrode equation (1913) which can be written in the following way:

$$\frac{S_t}{R} = \tfrac{5}{2} + 3 \ln \left(\frac{kT}{h} \frac{1}{\frac{v}{4}}\right) + \ln V \tag{2.4.8}$$

where V is the molecular volume. If the standard state is chosen to be that corresponding to one molecule per cm³, V is equal to unity and the standard translational entropy is given by:

$$\frac{S_t^0}{R} = \tfrac{5}{2} + 3 \ln \left(\frac{kT}{h} \frac{1}{\frac{v}{4}}\right) \tag{2.4.9}$$

In the case of rotation, for a rigid diatomic rotator AB, the entropy S_r can be written in the form:

$$\frac{S_r}{R} = 1 + 2 \ln \left(\frac{kT}{h} \frac{1}{\frac{v_\mu}{4}} \right) + \ln (4\pi\sigma^2) \tag{2.4.10}$$

where v_μ and σ have the same meaning as in (2.4.5). It must be noted that for internal motion, the notion of standard state is superfluous. If it is noted that the moment of inertia I of the molecule AB about an axis through its center of gravity and perpendicular to the line of centers, can be expressed in the form:

$$I = \mu\sigma^2 \tag{2.4.11}$$

it is easy to see that the entropy of rotation (2.4.10) can be rewritten in the more usual form:

$$\frac{S_r}{R} = 1 - \ln y \tag{2.4.12}$$

where

$$y = \frac{h^2}{8\pi^2 IkT} \tag{2.4.13}$$

The statistical mechanical expressions (2.4.9) and (2.4.10) for the translational and rotational entropies, are all that is required for the calculation of $\Delta S^{0\ddagger}$.

Indeed, because A and B are assumed to be hard spheres, they possess only entropy of translation. The "transition state" consisting of the spheres A and B in a state of collision possesses entropy of translation but also entropy of rotation. The transition state AB does not possess any vibrational entropy since its only vibrational mode is missing and, as remarked above, has been replaced by a translation along the collision coordinate. (Fig. 2.4.1).

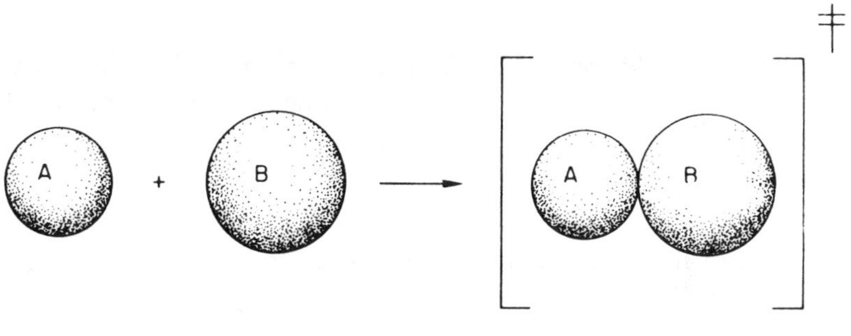

Fig 2.4.1 Collision between two hard spheres A and B.

Thus, the standard entropy of formation of the collision complex from collision partners is given by

$$\Delta S^{0\ddagger} = S^0_{t,\mathrm{AB}} + S_{r,\mathrm{AB}} - S^0_{t,\mathrm{A}} - S^0_{t,\mathrm{B}} \tag{2.4.14}$$

Substitution into (2.4.9) and (2.4.10) of the appropriate masses: $(m_\mathrm{A} + m_\mathrm{B})$ for the translation of AB, m_A and m_B for the translation of A and B respectively, as well as the expression (2.4.7) for μ, yields, with the help of (2.4.14):

$$\exp\left(\frac{\Delta S^{0\ddagger}}{R}\right) = \frac{h}{kT}\exp\left(-\frac{3}{2}\right)\pi\sigma^2 v_\mu \tag{2.4.15}$$

The second part of the problem now consists in the evaluation of the enthalpy of formation of AB from A and B. Since the system is a mixture of perfect gases, designation of a standard state for the enthalpy is superfluous so that superscripts zero can be omitted. Since the "reaction" between A and B considered here is a collision between hard spheres, there is no energy of interaction between the colliding partners and the energy of formation ΔE_0 at 0°K of the complex AB from A and B must be equal to zero:

$$\Delta E_0^{\ddagger} = 0 \tag{2.4.16}$$

Now, the relation between $\Delta E_0^{\ddagger}$ and $\Delta H^{\ddagger}$ can be obtained readily by noting first that

$$\Delta H^{\ddagger} = \Delta E^{\ddagger} - RT \tag{2.4.17}$$

since the "reaction" $\mathrm{A} + \mathrm{B} \rightarrow \mathrm{AB}$ is one where two moles of reactants give one mole of product. Moreover, the energy of formation $\Delta E^{\ddagger}$ at temperature T can be related to the energy of formation at 0°K by means of the equation of Kirchhoff:

$$\Delta E^{\ddagger} = \Delta E_0^{\ddagger} + \int_0^T \Delta C_v^{\ddagger} dT \tag{2.4.18}$$

where $\Delta C_v^{\ddagger}$ is the difference in molar heat capacities at constant volume between transition state and reactants. Since A and B are hard spheres and can translate only:

$$C_{v,\mathrm{A}} = C_{v,\mathrm{B}} = \tfrac{3}{2}R \tag{2.4.19}$$

On the other hand, the collision complex can both translate and rotate. Hence:

$$C_{v,\mathrm{AB}} = \tfrac{3}{2}R + R \tag{2.4.20}$$

From (2.4.19) and (2.4.20), it follows that

$$\Delta C_v^\ddagger = -\frac{R}{2} \tag{2.4.21}$$

Substitution of (2.4.21) into (2.4.18) yields:

$$\Delta E^\ddagger = \Delta E_0^\ddagger - \tfrac{1}{2}RT \tag{2.4.22}$$

Finally (2.4.16), (2.4.17) and (2.4.22) give:

$$\Delta H^\ddagger = -\tfrac{3}{2}RT \tag{2.4.23}$$

or

$$\exp\left(\frac{-\Delta H^\ddagger}{RT}\right) = \exp\left(\frac{3}{2}\right) \tag{2.4.24}$$

To identify the universal frequency ν as (kT/h), we now use (2.4.15) and (2.4.24) in (2.4.3). We get:

$$r = \nu \frac{h}{kT} \pi\sigma^2 v_\mu c_A c_B \tag{2.4.25}$$

Comparison with the well-known result (2.4.5) from the kinetic theory of gases completes the identification of the universal frequency as kT/h. We will therefore rewrite (2.4.3) as:

$$r = \frac{kT}{h} \exp\left(\frac{\Delta S^{0\ddagger}}{R}\right) \exp\left(\frac{-\Delta H^{0\ddagger}}{RT}\right) c_A c_B \tag{2.4.26}$$

This is the general equation of transition-state theory in its thermodynamic form. An equivalent form is

$$\boxed{r = \frac{kT}{h} K_c^\ddagger c_A c_B} \tag{2.4.27}$$

where use has been made of (2.4.2).

The thermodynamic formulation of rates is not particularly useful in calculations since standard entropies of activation $\Delta S^{0\ddagger}$ and standard enthalpies of activation $\Delta H^{0\ddagger}$ are rarely tabulated or calculated. But the formulation helps in understanding the nature of the problem of reaction rates. In order to have reaction, it is necessary to surmount not just an energy barrier but a *free-energy barrier*. If the energy barrier is low, the reaction path may well be rather improbable. In other words, if $\Delta E^{0\ddagger}$ is small, $\Delta S^{0\ddagger}$ may have a large negative value. Conversely, even though the energy barrier may be high, reaction may be helped if the entropy of activation is large and positive. These qualitative considerations indicate that *the energy factor and the entropy factor may tend to compensate each other*. They emphasize the fact that reaction is not a matter of energy alone but also requires reaching a favorable configuration accompanied by a change of entropy. They show the essential superiority of transition-state theory over the so-called *collision theory* where reaction is thought to occur if and only if collisions between reactants are sufficiently energetic. Yet, the collision theory remains enormously useful; it is intuitive, provides a norm from which abnormalities can be assessed, and deserves separate treatment.

2.5 The Collision Theory and the Equation of Arrhenius

According to the simple views of the collision theory of reaction rates, molecules are considered as hard spheres.

The rate of reaction between two such particles A and B is simply equal to the rate of collision with sufficient energy E to react. If we admit that the energy E is provided as relative translational energy along the line of centers of the colliding hard spheres, we obtain

$$r = \pi\sigma^2 v_\mu \exp\left(\frac{-E}{RT}\right) c_A c_B \qquad \text{(volume)} \qquad (2.5.1)$$

This expression is identical to the expression (2.4.5) except for the presence of the Boltzmann factor $\exp(-E/RT)$. This factor and the exact meaning of E just mentioned can be justified by the methods of the kinetic theory of gases. Similarly, for reaction of a hard sphere A at a surface with a minimum energy E required per collision:

$$r = \frac{v_A}{4} \exp\left(-\frac{E}{RT}\right) c_A \qquad \text{(surface)} \qquad (2.5.2)$$

Again, this is identical to (2.3.5) except for the presence of the Boltzmann factor. The mean molecular speed of A is denoted as v_A.

Comparison with transition-state theory can be made by introduction of a probability factor P expressing the fact that reactive collisions are not specified properly by gas kinetic theory, because the latter treats molecules as hard spheres:

$$r = P\,\pi\sigma^2 v_\mu \exp\left(-\frac{E}{RT}\right) c_A c_B \qquad \text{(volume)} \tag{2.5.3}$$

$$r = P\,\frac{v_A}{4} \exp\left(-\frac{E}{RT}\right) c_A \qquad \text{(surface)} \tag{2.5.4}$$

The task of transition-state theory is then, in principle, to calculate P *a priori* or to explain observed values of P in terms of particular features of the transition state. Because of the serious difficulties in even guessing the correct features of the transition state, it is almost invariably the second possibility that prevails, namely, P is explained *post facto*. It is already clear and will be explained presently that *the probability factor is closely related to the entropy of activation.* For reactions involving uncharged reactants, the probability factor is always less than unity. Thus the rates given by (2.5.1) and (2.5.2) have, in order of magnitude, their highest possible value. For the purpose of a rough estimate of the upper value of a rate of reaction, it is reasonable to take for the collision diameter σ a value based on viscosity diameters. This is not critical since an order of magnitude is all that is hoped for. Some very useful numbers based on commonly encountered values of the collision diameter, the molecular mass and the temperature are assembled in Table 2.5.1.

Table 2.5.1

USEFUL QUANTITIES IN CHEMICAL KINETICS

Quantity	Equation	Value	Order of Magnitude	Units
Mean molecular velocity v	(2.3.6)	$\left(\frac{8kT}{\pi m}\right)^{1/2}$	5×10^4	$\frac{\text{cm}}{\text{sec}}$
Universal frequency	(2.4.4)	$\frac{kT}{h}$	10^{13}	sec^{-1}
Collision frequency gas-surface	(2.3.5)	$\frac{v}{4}$	10^4	$\frac{\text{cm}}{\text{sec}}$
Collision frequency gas-gas	(2.4.5)	$\pi\sigma^2 v$	10^{-10}	$\frac{\text{cm}^3}{\text{sec}}$

It must be stressed that the usefulness of transition-state theory or of collision theory is *to provide a norm, not a number*. In applications, even a very rough value of P or of the entropy of activation can be of value in helping to eliminate a number of possibilities. Furthermore, in chemical kinetics, the data always come first. The purpose of the theory of rates of elementary steps is to serve as a guide in their interpretation.

The main results obtained thus far can be examined in terms of the rate constant k.

If there is only one reactant involved in an elementary step the rate is unimolecular and application of transition-state theory shows that the rate is also first-order.

Indeed, according to the general equation of transition-state theory (2.4.26) adapted to the case of a homogeneous reaction involving a single reactant A, we can write:

$$r = \frac{kT}{h} \exp\left(\frac{\Delta S^{0\ddagger}}{R}\right) \exp\left(\frac{-\Delta H^{0\ddagger}}{RT}\right) c_A \tag{2.5.5}$$

Thus the first-order rate constant k (a frequency, with a value given in $\sec^{-1}$) is:

$$k = \frac{kT}{h} \exp\left(\frac{\Delta S^{0\ddagger}}{R}\right) \exp\left(\frac{-\Delta H^{0\ddagger}}{RT}\right) \tag{2.5.6}$$

Considering that the temperature variation of $(kT/h) \exp (\Delta S^{0\ddagger}/R)$ can usually be neglected as compared to the strong exponential dependence of $\exp(-\Delta H^{0\ddagger}/RT)$, it is clear that (2.5.6) is, within an excellent approximation, of the Arrhenius form:

$$k = A \exp\left(-\frac{E}{RT}\right) \tag{2.5.7}$$

with

$$A = \frac{kT}{h} \exp\left(\frac{\Delta S^{0\ddagger}}{R}\right) \tag{2.5.8}$$

$$E = \Delta H^{0\ddagger} \tag{2.5.9}$$

In principle, as became clear in the preceding section, $\Delta H^{0\ddagger}$ depends on temperature, but only weakly so. Therefore, both A and E can be considered as independent of temperature.

Similarly, when the logarithm of an ordinary thermodynamic equilibrium constant is plotted versus reciprocal absolute temperature, a straight line is obtained with good approximation over an extended range of temperatures;

indeed, the temperature dependence of ΔS^0 and of ΔH^0 are negligible unless the temperature range becomes too wide.

It is expected that, in many cases, the transition state for a unimolecular reaction will have a structure resembling that of the reactant, except that one bond will be elongated prior to its rupture. If this resemblance in structure is close enough, as is very frequently the case, we expect that $\Delta S^{0\ddagger} \cong 0$ and therefore:

$$A \cong \frac{kT}{h} \tag{2.5.10}$$

Hence, the pre-exponential factor of the rate constant for unimolecular reaction is equal, in order of magnitude, to the universal frequency of transition-state theory. This conclusion is supported by a vast amount of experimental data. Exceptions to this rule can be ascribed to important changes of structure taking place in the transition state. It is still usually difficult to foresee such exceptions.

Consider next the case of a reactant A colliding with a surface. Following similar reasoning and using (2.5.4) and (2.5.5), we find, with similar approximations concerning the negligible dependence of pre-exponential factor A and activation energies E:

$$A = \frac{kT}{h} \exp\left(\frac{\Delta S^{0\ddagger}}{R}\right) = P\frac{v_{\mathrm{A}}}{4} \tag{2.5.11}$$

and again:

$$E = \Delta H^{0\ddagger} \tag{2.5.9}$$

Thus, the pre-exponential factor in the rate constant of a reaction following collision of a molecule with a surface has the dimensions of a linear velocity. The probability factor, which, as seen in a previous section, might be unity, for simple condensation of an atom on a surface, will in general be smaller than unity. It will be small if the reactant loses freedom upon reaching the transition state. A calculation of P would be tantamount to a calculation of $\Delta S^{0\ddagger}$.

Finally, if reaction between two reactants is considered, we find in similar fashion, using (2.4.26) and (2.5.3)

$$A = \frac{kT}{h} \exp\left(\frac{\Delta S^{0\ddagger}}{R}\right) = P\pi\sigma^2 v_\mu \tag{2.5.12}$$

$$E = \Delta H^{0\ddagger} \tag{2.5.9}$$

Here also, a small value of the probability factor will be associated with a large change in entropy upon formation of the transition state. This in turn will occur where the structure of the transition state differs substantially from that of the reactants. A useful guide on the necessity of considering such changes of structure is contained in a broad principle associated with Hammond.

Hammond's principle states essentially that if two states along the reaction coordinate have nearly the same energy, their molecular structures will resemble each other very closely and thus their entropies will also be closely related. This principle can be used, with caution, to guess values of the entropy of the transition state. Thus, if an elementary step is exothermic, the structure of the transition state is expected to resemble that of the reactants rather than that of the products. The reverse is true for an endothermic step.

But such principles, useful as they may be, are mere guides to chemical intuition and are of little help in quantitative calculations. Nevertheless, it will be stressed again that the general theory of rates of elementary steps sketched in this chapter is to provide norms of behavior and not numbers. In particular, transition-state theory provides a justification of the empirical laws of Guldberg and Waage and of Arrhenius. Much more importantly, by emphasizing the dual importance of entropy of activation and of enthalpy of activation, it provides a general framework for the accumulation and understanding of quantitative measurements of chemical reactivity. Finally, it gives a broad justification to the more intuitive concepts of collision theory, which remains enormously useful in qualitative considerations.

Problem 2.5.1

The standard entropy of a two-dimensional ideal gas (the standard state corresponding to a surface number density of unity) is given by:

$$S^0_{t,2} = 2 + 2\ln\left(\frac{kT}{h}\,\frac{1}{\frac{v}{4}}\right)$$

The entropy corresponding to rotation in a plane of a linear rotator AB may be taken simply as one half the value given by (2.4.10).

With this in mind, calculate the frequency of collisions of a gas with a surface, following a reasoning similar to that used in the previous section. Calculate also the rate of collisions between molecules A and B in a two-dimensional perfect gas mixture.

2.6 *Rates and Equilibrium*

An immediate important corollary of transition-state theory is that, *for an elementary step* in an ideal gas mixture:

$$A + B \rightleftarrows C + D$$

the ratio of rate constants must be equal to the thermodynamic equilibrium constant:

$$\boxed{\frac{\vec{k}}{\overleftarrow{k}} = K_c} \tag{2.6.1}$$

The expression for the net rate $r = \vec{r} - \overleftarrow{r}$ is:

$$r = \vec{k}c_A c_B - \overleftarrow{k}c_C c_D = \vec{r}\left(1 - \frac{1}{K_c}\frac{c_C c_D}{c_A c_B}\right) \tag{2.6.2}$$

Since from thermodynamics $-\Delta G = RT \ln (K_c/c_C c_D c_A{}^{-1} c_B{}^{-1})$,

$$r = \vec{r}(1 - e^{\Delta G/RT}) = \overleftarrow{r}(e^{-\Delta G/RT} - 1) \tag{2.6.3}$$

Sufficiently near equilibrium, $|\Delta G| \ll RT$ and

$$r = \frac{\vec{r}(-\Delta G)}{RT} = \frac{\overleftarrow{r}(-\Delta G)}{RT} = \frac{r_e(-\Delta G)}{RT} \tag{2.6.4}$$

since $\vec{r} = \overleftarrow{r} = r_e$ at equilibrium. Thus, under these conditions, the rate is proportional to the driving force for reaction: the free-energy difference with the minus sign, which is sometimes called the *affinity*.

Of course, at equilibrium $\Delta G = 0$ and $r = 0$.

The relation $r = r_e(-\Delta G/RT)$ can be put in a more explicit form. For the sake of generality, and for a reason that will become apparent in Chapter 4, let us assume that the reaction is of the general form:

$$0 = \sum_i \nu_i A_i \qquad i = 1, 2 \ldots, i, \ldots, c \tag{1.2.1}$$

Then, from thermodynamics, for a mixture of ideal gases:

$$-\Delta G = RT \ln \frac{(c_1)_e^{\nu_1} \cdots (c_c)_e^{\nu_c}}{(c_1)^{\nu_1} \cdots (c_c)^{\nu_c}} \tag{2.6.5}$$

where subscript e denotes concentrations at equilibrium.

In terms of number of moles n_i and $(n_i)_e$, this becomes

$$\Delta G = RT \sum_i \ln \left[\frac{n_i}{(n_i)_e}\right]^{\nu_i} = RT \sum_i \ln \left[1 + \frac{n_i - (n_i)_e}{(n_i)_e}\right]^{\nu_i} \tag{2.6.6}$$

Again, sufficiently near equilibrium: $\left|\frac{n_i - (n_i)_e}{(n_i)_e}\right| \ll 1$

Then

$$\Delta G = RT \sum_i \nu_i \frac{[n_i - (n_i)_e]}{(n_i)_e} \tag{2.6.7}$$

But

$$\left.\begin{aligned} n_i &= n_i^0 + \nu_i X \\ (n_i)_e &= n_i^0 + \nu_i X_e \end{aligned}\right\} \tag{1.2.2}$$

Consequently:

$$\Delta G = RT \sum_i \frac{\nu_i^2}{(n_i)_e} (X - X_e) \tag{2.6.8}$$

In terms of the extensive rate of reaction, the expression

$$R = \frac{dX}{dt} = R_e \frac{(-\Delta G)}{RT} \tag{2.6.9}$$

becomes:

$$\frac{dX}{dt} = R_e \sum_i \left[\frac{\nu_i^2}{(n_i)_e}\right](X_e - X) \tag{2.6.10}$$

or

$$\boxed{\frac{dX}{dt} = k(X_e - X)} \tag{2.6.11}$$

where

$$k = R_e \sum_i \frac{\nu_i^2}{(n_i)_e} \tag{2.6.12}$$

It is seen that k is a first-order rate constant, and that *sufficiently near equilibrium*, i.e., when $|\Delta G| \ll RT$ and $|n_i - (n_i)_e/n_i| \ll 1$, *all reactions are first-order: The rate is proportional to the distance away from equilibrium as measured by* $|X_e - X|$. The results of this section are valid only for an ideal system and an elementary step. In Chapter 4 the results will be generalized.

Problem 2.6.1

Demonstrate Eq. (2.6.1).

Problem 2.6.2

The rate constant of the elementary step

$$2\,HI \rightarrow H_2 + I + I$$

is 0.63 cm^3/g-mole-sec, at 698.6°K. What is the value of k for the reverse reaction

$$I + H_2 + I \rightarrow 2\,HI$$

at the same temperature? Obtain the necessary information from any handbook. [But see also J. H. Sullivan, *J. Chem. Phys.*, **46**, 73 (1967).]

2.7 *Rates in Solutions and in Thermodynamically Nonideal Systems*

All derivations so far have been restricted to the case of an elementary step in a mixture of ideal gases. The starting point was Eq. (2.3.2), stating that the rate is proportional to a frequency and to the concentration of transition states in equilibrium with reactants. This equation will now be rewritten in the form:

$$r = \frac{kT}{h} c_Z \tag{2.7.1}$$

which says that the rate is proportional to the concentration of transition states considered as a normal molecule except that motion along the reaction coordinate is now omitted from the description of this molecule. The constant of proportionality is now the universal frequency.

But in the case of a reaction in a system other than a mixture of ideal gases, (2.7.1) cannot be proved. Rather it will be assumed to hold, and experiment will decide whether the assumption is correct or not.

Previously, the concentration c_Z could be expressed simply in terms of the concentrations of reactants and of the equilibrium constant $K_c^\ddagger$. Indeed from (2.4.1):

$$\frac{c_Z}{c_A c_B} = K_c^\ddagger \tag{2.4.1}$$

But generally we do not expect (2.4.1) to hold true in the sense that the equilibrium constant is a constant only if activities a are used instead of concentrations c.

Thus, instead of (2.4.1), we must write:

$$\frac{a_Z}{a_A a_B} = K_c^\ddagger \tag{2.7.2}$$

Introducing activity coefficients γ such that $a = \gamma c$, we can write (2.4.1) in the form:

$$\frac{\gamma_Z}{\gamma_A \gamma_B} \frac{c_Z}{c_A c_B} = K_c^\ddagger \tag{2.7.3}$$

Solving (2.7.3) for c_Z and substituting into (2.7.1), we get:

$$r = \frac{kT}{h} K_c^\ddagger \frac{\gamma_A \gamma_B}{\gamma_Z} c_A c_B \tag{2.7.4}$$

The corresponding rate "constant" k is equal to:

$$k = \frac{kT}{h} K_c^\ddagger \frac{\gamma_A \gamma_B}{\gamma_Z} \tag{2.7.5}$$

At infinite dilution when the activity coefficients tend to unity, (2.7.4) reduces to (2.4.27) and the corresponding rate constant will now be called k_0 with a subscript to denote thermodynamic ideality:

$$k_0 = \frac{kT}{h} K_c^\ddagger \tag{2.7.6}$$

These considerations lead to a relation between the rate "constant" k in a nonideal system and the rate constant k_0 in the corresponding ideal reference system:

$$\boxed{\frac{k}{k_0} = \frac{\gamma_A \gamma_B}{\gamma_Z}} \tag{2.7.7}$$

This very important relation, originally due to Brönsted and Bjerrum, has been demonstrated beautifully in the case of dilute strong electrolytes for which the theory of Debye-Hückel applies. The theory gives an expression for the decimal logarithm of the activity coefficient γ_i of an ion of charge Z_i (positive or negative) in a dilute electrolyte solution characterized by an ionic strength $I = \frac{1}{2} \Sigma Z_i^2 c_i$ where c_i is the concentration (in g-mole/liter) of any strong electrolyte in solution:

$$-\log \gamma_i = c Z_i^2 \sqrt{I} \tag{2.7.8}$$

where c is a constant equal to 0.5 for solutions in water at 25°C.

Consider a reaction between A and B, ions of charges Z_A and Z_B respectively. Although as usual little may be known about the transition state, it must have a charge $(Z_A + Z_B)$. Substitution of the Debye-Hückel equation (2.7.8) into (2.7.7) gives:

$$\log\left(\frac{k}{k_0}\right) = 2cZ_AZ_B\sqrt{I} \tag{2.7.9}$$

All consequences of this relation have been verified both qualitatively and quantitatively. If one of the reactants is uncharged (say $Z_A = 0$), $k = k_0$. If both reactants are of the same charge, a plot of $\log(k/k_0)$ versus $\sqrt{I}$ is a straight line of positive slope. These correct conclusions would not be predicted by a naive approach that would write the following erroneous expression for the rate:

$$r = ka_Aa_B = k\gamma_A\gamma_Bc_Ac_B \tag{2.7.10}$$

Indeed, this would, by application of the Debye-Hückel expression, lead to

$$\log\left(\frac{k}{k_0}\right) = -c(Z_A^2 + Z_B^2)\sqrt{I} \tag{2.7.11}$$

Thus k would not be equal to k_0 if say Z_A were equal to zero. And the slope of $\log(k/k_0)$ versus $\sqrt{I}$ would always be negative. Furthermore relation (2.7.11) is also quantitatively wrong.

Thus the correct rate expression (2.7.4) to use in nonideal systems is known. Its application is however quite difficult at the present time because it is hard to guess at the proper value of the activity coefficient of the transition state even when activity coefficients of reactants are available. In practice, it will be necessary, regrettably, to forget about nonideality except under rare circumstances. The penalty for this simplification might be a rate "constant" that is not constant but drifts as the extent of reaction increases. Recourse to incorrect expressions such as (2.7.11) should be avoided.

Problem 2.7.1

In water solutions submitted to ionizing radiation, e.g., gamma rays, a very active intermediate is produced with a very short life. It has now been identified as a hydrated electron. This species can react with a variety of neutral and ionic species in solution. Devise an experiment which would provide a check on the electrical charge of the hydrated electron.

2.8 Summary

The dependence of the rate of elementary steps on temperature and composition follows directly from transition-state theory. But as is also the case normally in thermodynamics, the numerical value of the kinetic parameters, energy and entropy of activation, has to be determined experimentally. Nevertheless, theory provides a very useful guide in giving order-of-magnitude estimates of these parameters. The essential numerical results of this chapter are collected in Table 2.5.1. These results must be kept in mind at all times in kinetic work. Their usefulness will be demonstrated in the following chapters.

The thermodynamic formulation of reaction rates is also particularly useful in discussing rates in ideal solutions. Indeed, the concept of collision between molecules and the derivations of the kinetic theory of gases seem to be useless in the condensed state. Yet, the results of transition-state theory are not limited to the treatment of ideal gas mixtures. In particular, these results can also be couched in the language of the collision theory. This may appear surprising since the concept of collision in condensed phases is not a fruitful one. Yet it is found that "normal" reactions in solution exhibit a rate constant described by (2.5.3) with a probability factor P close to unity.

Reactions between uncharged species are said to be abnormally slow when P is much smaller than unity. An a priori calculation of the probability factor or of the standard entropy of activation is out of the question. But again, collision theory or transition-state theory define a normal behavior from which abnormalities can be assessed and understood a posteriori. In any event, theory is helpful only with respect to the pre-exponential factor of the rate constant. Just as is the case for reactions in ideal gases, the activation energy must always be obtained from experiment although it can sometimes be estimated in an empirical way (see Chapter 8) for the purpose of order-of-magnitude calculations.

BIBLIOGRAPHY

2.2 Codification of transition-state theory is largely the work of Eyring and his school. A general survey is presented in the first chapter of *The Theory of Rate Processes* by S. Glasstone, K. J. Laidler and H. Eyring, McGraw-Hill Book Company, Inc., New York, 1941.

2.4 All elementary demonstrations based on Eq. (2.4.1) lack rigor. For a more convincing approach which clarifies the fundamental assumption of equilibrium, see J. Ross and P. Mazur, *J. Chem. Phys.* **35**, 19 (1961).

The statistical mechanical results (2.4.9), (2.4.10), and (2.4.12) used in this chapter are presented in self-contained abbreviated form in *Thermodynamics* by G. N. Lewis and M. Randall, revised by K. S. Pitzer and L. Brewer, Chapter 27 and Appendix 3, McGraw-Hill Book Company, Inc., New York, 1961.

2.5 For an introduction to the subject, see R. H. Fowler and E. A. Guggenheim, *Statistical Thermodynamics*, Chapter 12, Cambridge University Press, New York, 1949.

2.7 The unsolved problems concerning the formulation of the theory of rate processes in solutions and thermodynamically nonideal systems have been reviewed critically by B. Kathleen Morse, "Investigation of Rates and Mechanisms of Reaction," Part I, Chapter XI, *Technique of Organic Chemistry* (2nd ed.), Vol. VIII, A. Weissberger, editor, Interscience Publishers, New York, 1961.

2.8 For a lucid introduction to some of the current problems in the theory and measurement of rates of elementary steps, see the very short paperback: *Modern Chemical Kinetics*, by Henry Eyring and Edward M. Eyring, Reinhold Publishing Corp., New York, 1963.

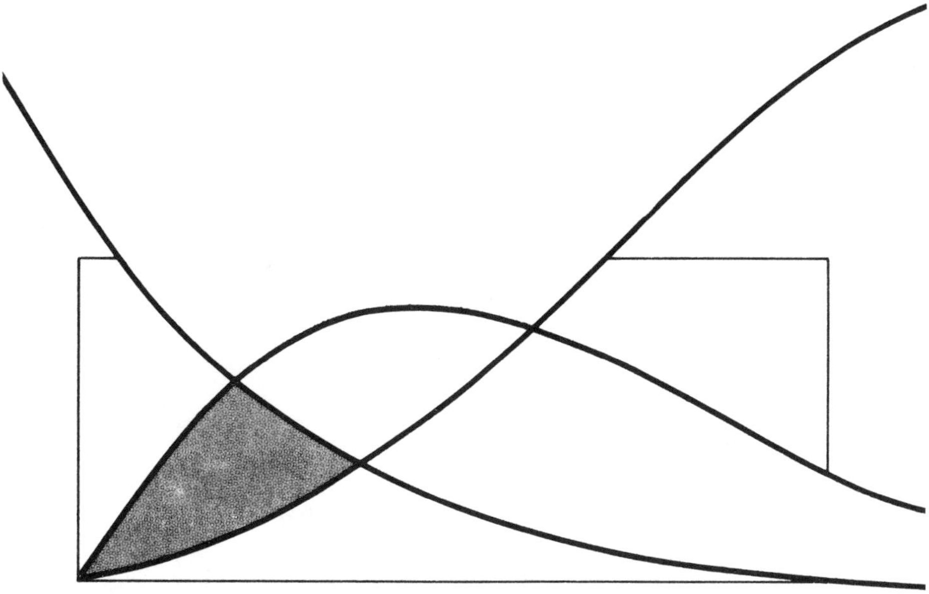

The Steady-State Approximation: Catalysis

3

3.1 *Single Reactions: Open and Closed Sequences*

One-step reactions between stable molecules are very rare. This is only natural since a stable molecule is, by definition, quite unreactive, and complicated rearrangements of bonds are usually required to go from reactants to products. This means that most directly observed reactions do not proceed in a single elementary step. Rather, a sequence of elementary steps is necessary, and reactants or products of these are not the stable reactants or products but are highly reactive intermediates that shall be called *active centers*.

Active centers may be of several different chemical types: free radicals, free ions, solvated ions, complexes at surfaces, complexes in a homogeneous phase, complexes with enzymes. Many active centers may be involved in a given reaction. Yet it is found that the advancement of the reaction can still be described by means of a single parameter — the extent of reaction. If this is the case, the reaction is said to be *single*. Why an apparently complex reaction remains stoichiometrically simple or single and how the kinetic treatment of such reactions can be handled by very general techniques are the two questions that will be handled in this chapter.

There are two distinct types of sequences leading from reactants to products through active centers. Sequences are either open or closed. An *open sequence* is one in which an active center is not reproduced in any other step of the sequence. A *closed sequence*, on the contrary, is one in which an active center is reproduced so that a cyclic reaction pattern repeats itself and a large number of molecules of products can be made from only one active center. A closed sequence is the best kinetic definition of *catalysis*.

In this sense the vast majority of reactions of theoretical or practical importance are catalytic, and the widespread interest in catalysis stems not only from academic but also from industrial reasons.

A few simple examples are listed in Table 3.1.1 to illustrate the terms active centers, open and closed sequences, and catalysis. For easy recognition, active centers are printed boldface. The stoichiometrically simple or single reaction is in each case obtained by summation of the elementary steps of the sequence, and this provides an easy check on the completeness of any proposed sequence.

Table 3.1.1

SEQUENCES AND ACTIVE CENTERS

Sequence	Type	Active Centers
$O_3 \rightarrow O_2 + \mathbf{O}$	open	oxygen atoms in the gas phase
$\mathbf{O} + O_3 \rightarrow O_2 + O_2$		
$2\,O_3 \rightarrow 3\,O_2$		
$RCl \rightarrow \mathbf{R^+} + Cl^-$	open	solvated (in liquid SO_2) benzhydryl carbonium ions: R^+
$\mathbf{R^+} + F^- \rightarrow RF$		
$RCl + F^- \rightarrow RF + Cl^-$		
$\mathbf{O} + N_2 \rightarrow NO + \mathbf{N}$	closed	oxygen and nitrogen atoms in the gas phase
$\mathbf{N} + O_2 \rightarrow NO + \mathbf{O}$		
$N_2 + O_2 \rightarrow 2\,NO$		

Table 3.1.1—*Cont.*

Sequence	Type	Active Centers
$\mathbf{H_3^+} + D_2 \rightarrow H_2 + \mathbf{HD_2^+}$	closed	gaseous molecular ions: H_3^+, HD_2^+, H_2D^+ and D_3^+
$\mathbf{HD_2^+} + D_2 \rightarrow HD + \mathbf{D_3^+}$		
$\mathbf{D_3^+} + H_2 \rightarrow D_2 + \mathbf{H_2D^+}$		
$\mathbf{H_2D^+ + H_2} \rightarrow \mathbf{HD + H_3^+}$		
$H_2 + D_2 \rightarrow 2\ HD$		
$\mathbf{SO_3^-} + O_2 \rightarrow \mathbf{SO_5^-}$	closed	free radical-ions SO_3^- and SO_5^- in water solution
$\mathbf{SO_5^-} + SO_3^{--} \rightarrow SO_5^{--} + \mathbf{SO_3^-}$		
$SO_5^{--} + SO_3^{--} \rightarrow 2\ SO_4^{--}$		
$2SO_3^{--} + O_2 \rightarrow 2SO_4^{--}$		
$\mathbf{S} + H_2O \rightarrow H_2 + \mathbf{SO}$	closed	sites S at the surface of a solid catalyst and surface complexes SO between S and an oxygen atom
$\mathbf{SO} + CO \rightarrow CO_2 + \mathbf{S}$		
$H_2O + CO \rightarrow H_2 + CO_2$		
$\mathbf{R} + O_2 \rightarrow \mathbf{ROO}$	closed	cumyl (R) and cumylperoxy (ROO) free radicals in solution in cumene (RH)
$\mathbf{ROO} + RH \rightarrow ROOH + \mathbf{R}$		
$RH + O_2 \rightarrow ROOH$		
$\mathbf{E} + CoH \rightleftarrows \mathbf{E{-}CoH}$	closed	an enzyme E and its complexes with a co-enzyme CoH and a substrate S
$\mathbf{E{-}CoH} + S \rightleftarrows \mathbf{E{-}CoH{-}S}$		
$\mathbf{E{-}CoH{-}S} \rightleftarrows \mathbf{E{-}Co{-}SH}$		
$\mathbf{E{-}Co{-}SH} \rightleftarrows \mathbf{E{-}Co + SH}$		
$\mathbf{E{-}Co} \rightleftarrows \mathbf{E} + Co$		
$CoH + S \rightleftarrows Co + SH$		

While all reactions showing a closed sequence may be said to be catalytic, there is a distinct difference between those where the active centers are provided by a separate entity called the *catalyst* which in principle has a very long lifetime and those where the active centers are generated within the system itself and may survive only during a limited number of cycles around the closed sequence. In the first category belong the truly *catalytic reactions*, in the narrow sense of the word. In the second category belong the *chain reactions*. Both types exhibit slightly different kinetic features. Yet they are so closely related that there is a great conceptual economy in considering them together. In particular, both types can be treated by means of the steady-state approximation which also applies to open sequences. This key development in chemical kinetics dates back to 1919 when J. A. Christiansen, K. F. Herzfeld and M. Polanyi proposed independently and almost at the same time a correct explanation of the kinetics of the reaction between hydrogen and bromine, reported in 1907 by M. Bodenstein and S. C. Lind.

3.2 The Steady-State Approximation

Consider a short open sequence consisting of two first-order irreversible elementary steps with rate constants k_1 and k_2:

$$\mathrm{A} \xrightarrow{k_1} \mathrm{B} \xrightarrow{k_2} \mathrm{C}$$

If (A_0) denotes the initial concentration of A at time $t = 0$ where also $(\mathrm{B}) = (\mathrm{C}) = 0$, the differential equations of the system are:

$$\frac{dx}{dt} = -k_1x, \qquad \frac{dy}{dt} = k_1x - k_2y \qquad \frac{dz}{dt} = k_2y \tag{3.2.1}$$

with $x = (\mathrm{A})/(\mathrm{A}_0)$; $y = (\mathrm{B})/(\mathrm{A}_0)$; $z = (\mathrm{C})/(\mathrm{A}_0)$. Integration gives simply:

$$x = \exp(-k_1t)$$

$$y = \frac{k_1}{k_2 - k_1}[\exp(-k_1t) - \exp(-k_2t)] \tag{3.2.2}$$

$$z = 1 - \frac{k_2}{k_2 - k_1}\exp(-k_1t) + \frac{k_1}{k_2 - k_1}\exp(-k_2t)$$

It is evident from (3.2.1) that

$$\frac{dx}{dt} + \frac{dy}{dt} + \frac{dz}{dt} = 0 \tag{3.2.3}$$

or

$$x + y + z = 1 \tag{3.2.4}$$

The concentration of A decreases monotonically. That of B first increases, reaches a maximum, then decreases. The maximum is reached at $t_{max} = [(1/(k_2 - k_1)] \ln (k_2/k_1)$ and the corresponding value of y is

$$y_{max} = \left(\frac{k_1}{k_2}\right)^{k_2/(k_2-k_1)} \tag{3.2.5}$$

At t_{max}, the curve of (C) versus t presents an inflection point, i.e., $d^2z/dt^2 = 0$. (Fig. 3.2.1).

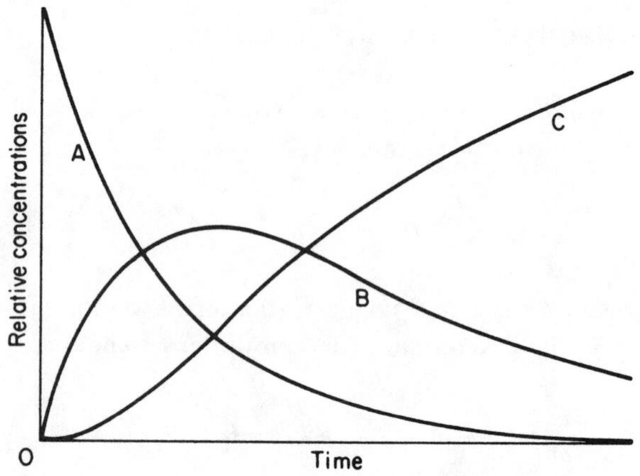

Fig. 3.2.1 Two first-order consecutive reactions

$$A \xrightarrow{k_1} B \xrightarrow{k_2} C$$

with $k_1 = 2k_2$. Note the zero slope of curve C at zero time, the maximum in curve B and the inflection point in curve C.

Suppose now that B is not an ordinary intermediate, but a very reactive one such as an active center. Kinetically this means that $k_2 \gg k_1$. What happens to the solution (3.2.2)? As $k_1/k_2 \to 0$, it reduces to:

$$x = \exp(-k_1 t) \qquad y = \frac{k_1}{k_2} \exp(-k_1 t) \qquad z = 1 - \exp(-k_1 t) \tag{3.2.6}$$

Also $t_{max} \to 0$ and $y_{max} \to 0$. Thus the time required for B to reach its maximum vanishingly small concentration is also vanishingly small. The inflection point in the curve of C is pushed back to the origin. The above equations (3.2.6) are the solutions of two differential equations and one algebraic equation:

$$-\frac{dx}{dt} = k_1x \qquad k_1x - k_2y = 0 \qquad \frac{dz}{dt} = k_2y \tag{3.2.7}$$

But the algebraic equation means that now:

$$\boxed{\frac{dy}{dt} = 0} \tag{3.2.8}$$

This is the analytical expression of the so-called steady-state approximation: *the derivative with respect to time of the concentration of active centers (reactive intermediates) is equal to zero.* The concentration of active centers is very small as compared to that of stable species.

But (3.2.8) must not be integrated: the result y = constant is obviously erroneous since the solution (3.2.6) shows that y decreases exponentially with time. What is significant is that B changes with time just as does A: $y = (k_1/k_2)x$, so that y does not depend explicitly on time but only implicitly through x, i.e., the concentration of a stable reactant.

Another equivalent statement of the steady-state approximation follows directly from (3.2.3).

Since $dy/dt = 0$, it follows that:

$$\frac{dx}{dt} = -\frac{dz}{dt} \tag{3.2.9}$$

Thus, *in a sequence of steps going through active centers as intermediates, the rates of reaction of the steps in the sequence are equal.* It also follows from (3.2.9) that *the reaction, however complex, can be described by a single parameter, the extent of reaction.* The reaction is written in the general form:

$$\sum \nu_i A_i = 0 \tag{1.2.1}$$

so that

$$\frac{dX}{dt} = \frac{1}{\nu_1}\frac{dn_1}{dt} = \cdots = \frac{1}{\nu_i}\frac{dn_i}{dt} \tag{3.2.10}$$

For the simple stoichiometry A → C, (3.2.10) simplifies to (3.2.9).

Of course, the steady-state approximation applies only after a time t_r, called the *relaxation time*, necessary for the steady-state concentration of the active centers to be approached. This relaxation time may be very small as compared to the total reaction time but although this situation will be frequently realized, there is no guarantee that this will be so, and in case of doubt a check is advisable.

Even past the relaxation time, the steady-state approximation remains an approximation but it will normally be completely satisfactory. To make these remarks more quantitative, it is of interest to inquire further into the nature of the approximation by returning to the case of the simple sequence A → B → C.

Let us assume that the actual concentration of B, namely (B), will differ from its steady-state approximation (B)*:

$$(\mathrm{B}) = (\mathrm{B})^*(1 - \epsilon) \tag{3.2.11}$$

This relation defines ϵ, the fractional deviation of the concentration of active center from its steady-state value.

As usual:

$$\frac{d(\mathrm{B})}{dt} = k_1(\mathrm{A}) - k_2(\mathrm{B}) \tag{3.2.12}$$

But also:

$$\frac{d(\mathrm{B})}{dt} = (\mathrm{B})^* \frac{d\epsilon}{dt} + (1 + \epsilon) \frac{d(\mathrm{B})^*}{dt} \tag{3.2.13}$$

According to the steady-state approximation:

$$(\mathrm{B})^* = \frac{k_1}{k_2} (\mathrm{A}) \tag{3.2.14}$$

Consequently, since $d(\mathrm{A})/dt = -k_1(\mathrm{A})$,

$$\frac{d(\mathrm{B})^*}{dt} = -\frac{k_1^2}{k_2} (\mathrm{A}) \tag{3.2.15}$$

Equating the right-hand sides of (3.2.12) and (3.2.13), and substituting the values for (B)* and $d(\mathrm{B})^*/dt$ from (3.2.14) and (3.2.15) we get:

$$\frac{d\epsilon}{dt} + (k_2 - k_1)\epsilon - k_1 = 0 \tag{3.2.16}$$

Integration, with initial condition $\epsilon = -1$ at $t = 0$, gives:

$$\epsilon = -\frac{1}{K - 1} [K - \exp (K - 1)\tau] \tag{3.2.17}$$

where $K = k_1/k_2$ and $\tau = k_2 t$.

Since B is supposed to be an active intermediate, $K = (k_1/k_2)$ must be smaller than unity. Then, at sufficiently large values of time:

$$\epsilon = \frac{K}{1 - K} \tag{3.2.18}$$

or if, as expected for an active center, $K \ll 1$,

$$\epsilon = K \tag{3.2.19}$$

for sufficiently large values of time. Thus the approximation is good if the ratio of rate constants is small and indeed ϵ, the measure of the error made in using the steady-state approximation, is equal to the ratio of rate constants.

What is meant by "sufficiently large values" of time is seen easily in the case of $K \ll 1$. Then the expression for ϵ reduces to:

$$\epsilon = -e^{-\tau} \tag{3.2.20}$$

The relaxation time, in kinetics, is defined as the time required for a quantity to decay to a fraction $1/e$ of its original value. Thus the relaxation time, in this case, is defined by $\tau = 1$ or $t_r = 1/k_2$. The relaxation time, for the simple sequence, is thus equal to the inverse of its first-order rate constant for reaction, or to the average lifetime of the active center.

Unfortunately, there exists no general theory that does for a general sequence of elementary steps what has been done here for the simple sequence of first-order reactions. Yet the general ideas are clear. While exceptions to the validity of the steady-state approximation are known, they are rare and *the steady-state approximation can be considered as the most important general technique of applied chemical kinetics*. The treatment of long sequences becomes a simple problem as will now be shown.

Problem 3.2.1

For the sequence A $\rightarrow$ B $\rightarrow$ C of two first-order reactions, calculate the maximum concentration of B and the time required to reach this maximum, when both rate constants are equal.

3.3 Kinetic Treatment of Catalytic Sequences

Consider the reaction

$$A_1 + A_2 + A_3 + A_4 \rightleftarrows B_1 + B_2 + B_3 + B_4$$

going through the following sequence involving the active centers $\mathbf{X}_1$, $\mathbf{X}_2$, $\mathbf{X}_3$, and $\mathbf{X}_4$:

$$A_1 + \mathbf{X}_1 \rightleftarrows \mathbf{X}_2 + B_2 \qquad (1)$$

$$A_2 + \mathbf{X}_2 \rightleftarrows \mathbf{X}_3 + B_3 \qquad (2)$$

$$A_3 + \mathbf{X}_3 \rightleftarrows \mathbf{X}_4 + B_4 \qquad (3)$$

$$A_4 + \mathbf{X}_4 \rightleftarrows \mathbf{X}_1 + B_1 \qquad (4)$$

The sequence is closed. It is a simple one in the sense that each elementary step is first-order with respect to an active center in both directions. Rate constants for the ith step are k_i from left to right and k_{-i} from right to left. It is convenient to use the notation $a_i = k_i(A_i)$ and $a_{-i} = k_{-i}(B_i)$. The a's are therefore pseudo first-order rate constants, e.g., a_i is 0.69 times the inverse of the half-life of $\mathbf{X}_i$ in step (1) from left to right [see (1.6.14)]. The net rate of each step is r_i. According to the steady-state approximation:

$$r_1 = r_2 = r_3 = r_4 = r \qquad (3.3.1)$$

The equations of the problem are the four algebraic equations:

$$\begin{aligned} a_1(\mathbf{X}_1) - a_{-1}(\mathbf{X}_2) &= r \\ a_2(\mathbf{X}_2) - a_{-2}(\mathbf{X}_3) &= r \\ a_3(\mathbf{X}_3) - a_{-3}(\mathbf{X}_4) &= r \\ a_4(\mathbf{X}_4) - a_{-4}(\mathbf{X}_1) &= r \end{aligned} \qquad (3.3.2)$$

The unknowns are $(\mathbf{X}_i)/r$ for $i = 1,2,3,4$:

$$\begin{aligned} \frac{a_1(\mathbf{X}_1)}{r} - \frac{a_{-1}(\mathbf{X}_2)}{r} &= 1 \\ \frac{a_2(\mathbf{X}_2)}{r} - \frac{a_{-2}(\mathbf{X}_3)}{r} &= 1 \\ \frac{a_3(\mathbf{X}_3)}{r} - \frac{a_{-3}(\mathbf{X}_4)}{r} &= 1 \\ \frac{a_4(\mathbf{X}_4)}{r} - \frac{a_{-4}(\mathbf{X}_1)}{r} &= 1 \end{aligned} \qquad (3.3.3)$$

Following Christiansen, the solution can be written in the following convenient form:

$$\frac{(\mathbf{X}_i)}{r} = \frac{M_i}{\Delta} \qquad (i = 1,2,3,4) \tag{3.3.4}$$

where

$$\Delta \equiv a_1a_2a_3a_4 - a_{-1}a_{-2}a_{-3}a_{-4} \tag{3.3.5}$$

and M_i is the sum of the elements of the ith row of a square matrix M of i rows and i columns, each element being a product of $(i - 1)$ different a's:

$$M \equiv \begin{vmatrix} a_2a_3a_4 & a_{-1}a_3a_4 & a_{-1}a_{-2}a_4 & a_{-1}a_{-2}a_{-3} \\ a_3a_4a_1 & a_{-2}a_4a_1 & a_{-2}a_{-3}a_1 & a_{-2}a_{-3}a_{-4} \\ a_4a_1a_2 & a_{-3}a_1a_2 & a_{-3}a_{-4}a_2 & a_{-3}a_{-4}a_{-1} \\ a_1a_2a_3 & a_{-4}a_2a_3 & a_{-4}a_{-1}a_3 & a_{-4}a_{1-}a_{-2} \end{vmatrix} \tag{3.3.6}$$

This matrix is easy to construct by alternation of subscripts if it is built column by column, the first element being $a_2a_3a_4$. Now if $(\mathbf{L})$ is the concentration of catalyst in the system (e.g., number of active surface sites per unit area of solid or number of active enzymatic groups per unit volume):

$$(\mathbf{L}) = \sum_i (\mathbf{X}_i) \tag{3.3.7}$$

Consequently

$$\sum_i \frac{(\mathbf{X}_i)}{r} = \frac{1}{\Delta}\sum_i M_i = \frac{(\mathbf{L})}{r} \tag{3.3.8}$$

and finally:

$$\boxed{r = (\mathbf{L}) \frac{\Delta}{\sum_i M_i}} \tag{3.3.9}$$

The rate is proportional to the catalyst concentration and to an expression (3.3.5):

$$\Delta = k_1k_2k_3k_4(\mathrm{A}_1)(\mathrm{A}_2)(\mathrm{A}_3)(\mathrm{A}_4) - k_{-1}k_{-2}k_{-3}k_{-4}(\mathrm{B}_1)(\mathrm{B}_2)(\mathrm{B}_3)(\mathrm{B}_4) \tag{3.3.10}$$

of the form of the mass action law of Guldberg and Waage. But it is also inversely proportional to a sum $\sum_i M_i$ of terms. Each of these i^2 terms is the product of $(i - 1)$ rate constants and $(i - 1)$ concentrations of stable reactants or products.

This result is very important: it indicates the general functional form of the rate r for a stoichiometrically simple reaction. But, in practice, the ponderous solution (3.3.9) will be arrived at, not by blind application of the formulae involving the matrix (3.3.6), but by applying the general condition for the steady state (3.2.8) to every active center and by noting that, with a catalyst of constant activity, the sum of concentrations of active centers is equal to a constant.

Thus, in the case of a single reaction:

$$A_1 + A_2 \rightleftarrows B_1 + B_2$$

taking place through a two-step catalytic sequence:

$$A_1 + \mathbf{X}_1 \rightleftarrows \mathbf{X}_2 + B_1 \qquad (1)$$

$$A_2 + \mathbf{X}_2 \rightleftarrows \mathbf{X}_1 + B_2 \qquad (2)$$

we can write, with the same notations as above:

$$\frac{d(\mathbf{X}_1)}{dt} = -a_1(\mathbf{X}_1) + a_{-1}(\mathbf{X}_2) + a_2(\mathbf{X}_2) - a_{-2}(\mathbf{X}_1) = 0 \qquad (3.3.11)$$

$$\frac{d(\mathbf{X}_2)}{dt} = a_1(\mathbf{X}_1) - a_{-1}(\mathbf{X}_2) - a_2(\mathbf{X}_2) + a_{-2}(\mathbf{X}_1) = 0 \qquad (3.3.12)$$

$$(\mathbf{X}_1) + (\mathbf{X}_2) = (\mathbf{L}) \qquad (3.3.13)$$

It is clear that (3.3.11) and (3.3.12) are identical:

$$(\mathbf{X}_1)(a_1 + a_{-2}) = (\mathbf{X}_2)(a_{-1} + a_2) \qquad (3.3.14)$$

Thus the ratio of the concentrations of active centers is given by:

$$\frac{(\mathbf{X}_1)}{(\mathbf{X}_2)} = \frac{a_{-1} + a_2}{a_1 + a_{-2}} \qquad (3.3.15)$$

Hence:

$$\frac{(\mathbf{X}_1)}{(\mathbf{X}_1) + (\mathbf{X}_2)} = \frac{(\mathbf{X}_1)}{(\mathbf{L})} = \frac{a_{-1} + a_2}{a_1 + a_2 + a_{-1} + a_{-2}} \qquad (3.3.16)$$

$$(\mathbf{X}_2) = (\mathbf{L}) \frac{a_1 + a_{-2}}{a_1 + a_2 + a_{-1} + a_{-2}} \qquad (3.3.17)$$

Since the rate of the reaction is equal to that of either step:

$$r = r_1 = a_1(\mathbf{X}_1) - a_{-1}(\mathbf{X}_2) = (\mathbf{L})\,\frac{a_1 a_2 - a_{-1}a_{-2}}{a_1 + a_2 + a_{-1} + a_{-2}} \qquad (3.3.18)$$

This result is, of course, the same as that obtained by direct application of the formula (3.3.9).

Problem 3.3.1

Find the rate of the water-gas shift reaction

$$CO + H_2O \rightleftarrows CO_2 + H_2$$

taking place through a two-step sequence as shown in Table 3.1.1, at the surface of a solid catalyst. What physical meaning could be attached to a rate expression of the form

$$r = \frac{k(CO)(H_2O)}{(CO_2) + B(H_2O)}$$

where k and B are constants at a given temperature?

Problem 3.3.2

It is assumed that the reaction

$$2\ SO_2 + O_2 \rightleftarrows 2\ SO_3$$

takes place at the surface of a solid catalyst through the three-step sequence:

$$\mathbf{S} + O_2 \rightleftarrows \mathbf{S{-}O{-}O}$$

$$\mathbf{S{-}O{-}O} + SO_2 \rightleftarrows SO_3 + \mathbf{S{-}O}$$

$$\mathbf{S{-}O} + SO_2 \rightleftarrows SO_3 + \mathbf{S}$$

Find the rate expression.

3.4 Kinetic Treatment of Chain Reactions

A chain reaction is a closed catalytic sequence but the active centers are not provided by a catalyst. Rather they are generated in the system by reaction of a molecule called the *initiator*, e.g., ditertiary-butyl peroxide, or by electromagnetic radiation (e.g., UV light), or by ionizing radiation (e.g., gamma rays). The steps of the closed sequence are *propagation* steps and the rate r determined in the previous section is the rate of propagation. But besides, the closed sequence has one or several heads and tails. A head is a step that produces active centers. The rate of production of active centers is the rate of *initiation:* r_i. A tail is a step that destroys active centers. The rate of destruction of active centers is the rate of *termination:* r_t. The *kinetic chain length* or simply chain length is defined as the ratio r/r_i. If it is sufficiently larger than unity, i.e., in the case of long chains, the by-products that may be the result of the head and tail reactions will represent only a very small fraction of the products of the propagation steps. Thus within an approximation that improves with chain length, the reaction remains stoichiometrically simple. This simplification of the problem is a result of what is frequently called the *long-chain approximation.*

At the steady state, the rate of initiation must be equal to the rate of termination, otherwise the concentration of active centers would depend explicitly on time. The relation

$$r_i = r_t \tag{3.4.1}$$

provides one additional relation to eliminate the unknown concentration of one of the active centers propagating the chain, say $(\mathbf{X}_1)$. Thus the rate expression can be found from (3.3.4):

$$\frac{(\mathbf{X}_1)}{r} = \frac{M_1}{\Delta} \tag{3.4.2}$$

together with $r_i = r_t$.

If there is only one head and one tail and these are identical, the equality $r_i = r_t$ means that the concentration of the particular active center involved, say $\mathbf{X}_i$, is an equilibrium concentration at the steady state:

$$(\mathbf{X}_i)^* = (\mathbf{X}_i)_e \tag{3.4.3}$$

In general, of course, this is not true and steady-state concentrations $(\mathbf{X}_i)^*$ are different from equilibrium concentrations $(\mathbf{X}_i)_e$.

The final form of the rate equation will be dictated in large measure by the detail of initiation and termination steps. Inclusion of all possibilities frequently leads to intractable results. To retain only the significant steps is somewhat of an art based on information available on the rate constants of these elementary steps. Frequently mere order-of-magnitude estimates of these rate constants will be very helpful in excluding a number of possibilities and obtaining a tractable rate equation which can be confronted with available kinetic data. These special techniques which rely on a simplification of the general problem will be illustrated in the next chapter.

The general rate expressions developed in this section appear quite formidable. Yet, frequently, if the sequence is a simple one and if there is only one head and one tail, the final rate expression is surprisingly simple: first-order, second-order, or fractional-order. Again, as in the case of catalytic sequences, the rate equation is best derived by systematic application of the steady-state conditions (3.2.8) rather than by routine use of Eqs. (3.4.1) and (3.4.2).

Problem 3.4.1

Derive the rate expression for the reaction:

$$N_2 + O_2 \rightleftarrows 2\,NO$$

proceeding through the two-step closed sequence shown in Table 3.1.1. It is further assumed that oxygen atoms are in equilibrium with oxygen molecules, i.e., the rate of dissociation of molecular oxygen into atoms (initiation, head) is equal to the rate of recombination of oxygen atoms (termination, tail).

Problem 3.4.2

Derive the rate expression for the reaction:

$$H_2 + Br_2 \rightarrow 2\,HBr$$

The closed sequence is:

$$\mathbf{Br} + H_2 \rightleftarrows HBr + \mathbf{H}$$

$$\mathbf{H} + Br_2 \rightarrow HBr + \mathbf{Br}$$

Initiation is decomposition of bromine molecules and termination is recombination of bromine atoms. Note that the first step of the sequence is assumed to be reversible but not the second.

Problem 3.4.3

Derive the rate expression for the reaction:

$$C_2H_4 + H_2 \rightarrow C_2H_6$$

taking place through the closed sequence:

$$\mathbf{H} + C_2H_4 \rightarrow \mathbf{C_2H_5}$$

$$\mathbf{C_2H_5} + H_2 \rightarrow C_2H_6 + \mathbf{H}$$

with initiation:

$$C_2H_4 + H_2 \rightarrow \mathbf{C_2H_5} + \mathbf{H}$$

and termination:

$$\mathbf{C_2H_5} + \mathbf{H} \rightarrow C_2H_6$$

Problem 3.4.4

Derive the rate expression for the reaction:

$$CH_3CHO \rightarrow CH_4 + CO$$

taking place through the closed sequence:

$$\mathbf{CH_3CO} \rightarrow \mathbf{CH_3} + CO$$
$$\mathbf{CH_3} + CH_3CHO \rightarrow \mathbf{CH_3CO} + CH_4$$

with initiation:

$$CH_3CHO \rightarrow \mathbf{CH_3} + \mathbf{CHO}$$

and termination:

$$\mathbf{CH_3} + \mathbf{CH_3} \rightarrow C_2H_6$$

It is assumed that the chains are long.

3.5 *Principle of Microscopic Reversibility*

Consider the sequence of elementary steps:

$$\mathbf{X}_1 \rightleftarrows \mathbf{X}_2 \qquad (1)$$
$$\mathbf{X}_2 \rightleftarrows \mathbf{X}_3 \qquad (2)$$
$$\mathbf{X}_3 \rightleftarrows \mathbf{X}_1 \qquad (3)$$

which represents a simple triangular network of the form shown on page 3.

At the steady state, no net chemical change takes place, so that at the steady state, the system is also in a state of chemical equilibrium:

$$(\mathbf{X}_i)^* = (\mathbf{X}_i)_e \qquad (3.5.1)$$

Applying the equations of Section 3 to this simple closed sequence, and noting that the *a*'s are identical to the *k*'s for the present situation, we have:

$$\Delta = k_1k_2k_3 - k_{-1}k_{-2}k_{-3} \tag{3.5.2}$$

$$\mathrm{M} \equiv \begin{vmatrix} k_2k_3 & k_{-1}k_3 & k_{-1}k_2 \\ k_3k_1 & k_{-2}k_1 & k_{-2}k_{-3} \\ k_1k_2 & k_{-3}k_2 & k_{-3}k_{-1} \end{vmatrix} \tag{3.5.3}$$

Thus the rate at the steady state is given by:

$$r = (\mathbf{X}_1)^* \frac{\Delta}{M_1} \tag{3.5.4.}$$

and the ratio of the concentrations of $\mathbf{X}_1$ and $\mathbf{X}_2$ at the steady state is:

$$\frac{(\mathbf{X}_2)^*}{(\mathbf{X}_1)^*} = \frac{k_3k_1 + k_{-2}k_1 + k_{-2}k_{-3}}{k_2k_3 + k_{-1}k_3 + k_{-1}k_{-2}} \tag{3.5.5}$$

But according to (3.5.1), $(\mathbf{X}_2)^* = (\mathbf{X}_2)_e$ and $(\mathbf{X}_1)^* = (\mathbf{X}_1)_e$. Consequently, since the ratio of concentrations at equilibrium must be equal to the equilibrium constant K_1 of the first step, we obtain for K_1 the expression:

$$K_1 = \frac{k_3k_1 + k_{-2}k_1 + k_{-2}k_{-3}}{k_2k_3 + k_{-1}k_3 + k_{-1}k_{-2}} \tag{3.5.6}$$

As can easily be verified, this expression reduces to the one that is known to prevail for a single elementary step:

$$K_1 = \frac{k_1}{k_{-1}} \tag{2.9.1}$$

only if:

$$k_1k_2k_3 = k_{-1}k_{-2}k_{-3} \tag{3.5.7}$$

which, following (3.5.2) and (3.5.4), means $\Delta = 0$ and $r = 0$. Thus, if it is assumed that at equilibrium the net rate is equal to zero, we find that, even in a system of interconnected elementary steps, we still have:

$$\boxed{K_i = \frac{k_i}{k_{-i}}} \tag{3.5.8}$$

Therefore, at equilibrium, any given elementary step and its reverse proceed at the same rate. This conclusion is the expression of the *principle of detailed balancing or microscopic reversibility or entire equilibrium.* It is an additional principle because from usual thermodynamic and kinetic relations alone, the equilibrium situation could be represented by a different relation, and an additional condition, namely $r = 0$, had to be imposed before the expression for detailed balancing could be written down.

These considerations have played an important role in the origin of the thermodynamics of irreversible processes. The principle of detailed balancing provides an automatic check on the self-consistency of postulated reaction sequences when equilibrium is approached from opposite sides. It will be used more specifically in the treatment of reaction networks in Chapter 10.

3.6 Kinetics of Polymerization

In the polymerization of monomers to materials of high molecular weight, the product is a petrified sample of the fleeting kinetics that governed its synthesis. For example, the polymer chain is the materialization of the kinetic chain inferred in many other chemical changes. The polymerization of vinyl monomers in particular is a beautiful chapter in chemical kinetics and its study is very rewarding. It is also a very large chapter and in this section, the subject is introduced only as an illustration of general kinetic principle.

In addition polymerization, a molecule of monomer M is transformed into an active center **M** by reaction with an active center produced normally by decomposition of a molecule of initiator I. In turn, **M** reacts with another monomer molecule to produce a dimer active center. This propagation continues and the growing polymer chain represents an active center which will be designated by **M** without reference to its size, called the *degree of polymerization*, $\overline{\mathrm{DP}}$, i.e., the number of monomeric units in a chain of polymer. This is permissible because the rate of propagation does not depend on the size of the growing active center.

The elementary steps of addition polymerization are then:

$$\left.\begin{array}{rcll} \mathrm{I} & \rightarrow & \mathbf{R} + \mathbf{R} & \text{(i)} \\ \mathbf{R} + \mathrm{M} & \rightarrow & \mathbf{M} & \\ \mathbf{M} + \mathrm{M} & \rightarrow & \mathbf{M} & \text{(p)} \\ \mathbf{M} + \mathbf{M} & \rightarrow & \mathrm{M}_2 & \text{(t)} \end{array}\right\} \quad (3.6.1)$$

The second step is not kinetically significant. To fix the ideas, it has been assumed that one molecule of initiator decomposes thermally into two active centers **R** that yield the active centers **M**. Also, two growing chains terminate the sequence by mutual recombination. The rate of polymerization, r, i.e., the rate of disappearance of monomer, i.e., the rate of propagation if, as expected, the kinetic chain length is sufficiently long, is given by:

$$r = a_p(\mathbf{M})$$

But, at steady state, $r_i = a_i = k_i(\mathbf{M})^2 = r_t$. Consequently:

$$\boxed{r = a_p\left(\frac{a_i}{k_t}\right)^{1/2} = k_p\left(\frac{k_i}{k_t}\right)^{1/2}(\mathrm{I})^{1/2}(\mathrm{M})} \qquad (3.6.2)$$

This rate expression obtains in the case of the ideal sequence (3.6.1). Naturally many variations and complications are possible and are actually observed. In the case at hand, the average degree of polymerization $\overline{\mathrm{DP}}$ is equal to twice the kinetic chain length. The factor two arises because termination occurs through recombination of two growing chains. The degree of polymerization is an average because so is the kinetic chain length; it is the number of times, on the average, that a chain carrier goes around the closed sequence before it is destroyed.

Suppose now there exists in the reacting mixture a molecule S which can react with **M** to become an active center **S**. The latter is then able to start a new chain:

$$\mathbf{M} + \mathrm{S} \rightarrow \mathbf{S} + \text{dead chain (tr)}$$

$$\mathbf{S} + \mathrm{M} \rightarrow \mathbf{M}$$

This substance S may be added on purpose or it may be an impurity. If the active center **S** is sufficiently reactive, the presence of S will not depress the rate of polymerization or the chain length. But it will decrease the degree of polymerization since the growth of the polymer chain is interrupted in the reaction between **M** and S, which is therefore called a *chain transfer* process. The efficiency of the chain transfer is determined by a selectivity constant $c_S = (k_{tr}/k_p)$ which can be determined easily in principle. Indeed

$$-\frac{d(\mathrm{S})}{dt} = k_{tr}(\mathbf{M})(\mathrm{S}) \qquad -\frac{d(\mathrm{M})}{dt} = k_p(\mathbf{M})(\mathrm{M})$$

Dividing side by side, we get:

$$\frac{d(\mathrm{S})}{d(\mathrm{M})} = c_S\frac{(\mathrm{S})}{(\mathrm{M})}$$

or after integration:

$$\boxed{\frac{\log (\mathrm{S})_0 - \log (\mathrm{S})}{\log (\mathrm{M})_0 - \log (\mathrm{M})} = c_\mathrm{S}} \tag{3.6.3}$$

Thus c_S can be determined by measuring the concentrations of S and M at time zero and any time t.

The applicability of the steady-state approximation that led to the standard rate expression (3.6.2) for polymerization is normally unquestioned.

However, the study of the kinetics of polymerization under nonsteady-state conditions can lead to the separate determination of the mean lifetime of the active centers. This can be done when the polymerization is initiated by light of intensity I such that

$$r_i = k_i I$$

For the same type of termination, the rate equation for polymerization is then identical to (3.6.2). In particular the rate of polymerization is proportional to the square root of the light intensity.

Suppose now that the initiating light beam is directed to the reactor through a rotating slotted disk. The reacting system will be subjected to alternating periods of light and darkness and for simplicity, let us assume that these periods are of equal duration.

If, first of all, the duration of a period of light or darkness is very much larger than the relaxation time required to reach the steady state, the reactor will be essentially submitted to the intensity I for half of the time and the rate will be proportional to $I^{1/2}/2$.

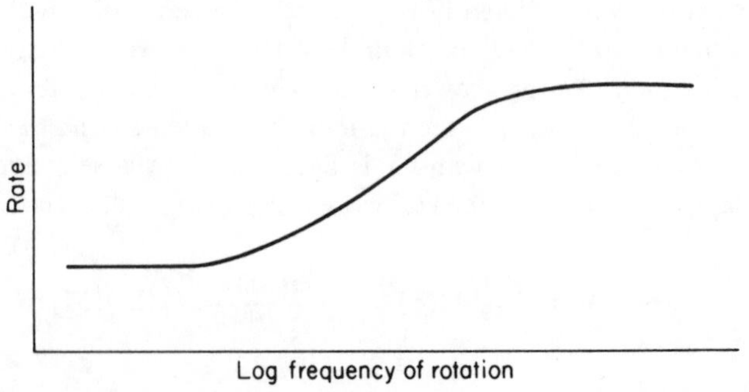

Fig. 3.6.1 Rate of polymerization. A typical dispersion curve obtained by intermittent illumination by means of a rotating sector located between the source of light and the reacting medium.

If, on the other hand, the speed of rotation of the sector is increased so that the duration of a period of light or darkness becomes much smaller than the relaxation time of the active centers, everything happens as if the reactor were illuminated all the time by half the intensity I and the rate will now be proportional to $(I/2)^{1/2}$ and is therefore $2^{1/2}$ times larger than at low sector speed. If the rate of reaction is plotted versus the logarithm of the frequency of rotation, a "dispersion wave" will be obtained (Fig. 3.6.1) and the inflection point of the wave corresponds to the mean lifetime of the active centers.

This is an example where departure from the steady-state condition can be produced deliberately in order to reach a useful kinetic goal. But, in general, if there are reasons to believe that the steady-state approximation does not apply, the situation is usually very bad. Thus, in addition polymerization, while the steady-state approximation normally applies to the total concentration of active centers, it is much more questionable to apply the steady-state condition to individual classes of active centers, i.e., active polymer chains of a given length. Yet, to obtain a distribution of molecular weights, the steady-state approximation, as applied to active polymer chains, is the only technique leading to tractable results. Without it, calculations become quite formidable. It may be noted that the distribution of molecular weights in the polymeric product can be shown to be sensitive to the mode of termination and even to the type of reaction system chosen. Thus different types of distribution are obtained in batch and in stirred-flow reactors. Since the molecular weight distribution affects the properties of the polymer, it is clear that a number of basic kinetic principles are of decisive importance in polymer chemistry — the steady-state approximation and the type of reactor in particular.

Problem 3.6.1

Show that the chain transfer constant c_S can also be determined by measuring the average degree of polymerization both in the absence $[\overline{DP}]_0$ and in the presence $[\overline{DP}]$ of chain transfer agent:

$$\frac{1}{[\overline{DP}]} = \frac{1}{[\overline{DP}]_0} + c_S \frac{(S)}{(M)}$$

This expression shows the effect of S on the polymer chain length.

BIBLIOGRAPHY

3.1 Although the term "active center" is frequently associated with reactive sites at a solid surface following the proposal of Hugh S. Taylor in 1925, the name has also been used extensively, especially by Semenov's school, to denote reactive intermediates in chain reactions. Other names are "reaction centers," and "chain centers." (F. S. Dainton, *Chain Reactions*, Methuen & Co. Ltd., London 1956). As it is felt desirable to emphasize the common mode of kinetic action of all catalysts, the term "active center" is used here for all active intermediates, irrespective of their nature.

3.2 The steady-state approximation, advocated by Bodenstein in the 1920's, is now a classical tool of chemical kinetics. Its limitations, as sketched in this section, are still under discussion. The present treatment is based on the work of J. C. Giddings and H. K. Shin, *Trans. Faraday Soc.*, **57**, 468 (1961). A more powerful approach has been outlined by J. R. Bowen, A. Acrivos and A. K. Oppenheim, *Chem. Eng. Sci.* **18**, 177 (1963).

3.3 The kinetic treatment of sequences has been systematized by Christiansen, from whose review in *Advan. Catalysis*, **5**, 311 (1953), much of this section is borrowed.

3.4 The long chain approximation in free radical reaction systems has been examined more recently by G. R. Gavalas, *Chem. Eng. Sci.*, **21**, 133 (1966).

3.5 The principle of microscopic reversibility is of a more fundamental character than indicated here. But implicit use of the principle was made by the chemical kineticist, as explained in this section, prior to the work of Onsager, who actually was led to his Reciprocal Relations by noting the balancing of individual reactions in a reaction network. See K. G. Denbigh, *The Thermodynamics of the Steady State*, Methuen & Co. Ltd., London, 1951, for an elementary introduction to the subject.

3.6 For further details, see *The Kinetics of Vinyl Polymerization by Radical Mechanisms* by C. H. Bamford, W. G. Barb, A. D. Jenkins, and P. F. Onyon, Academic Press Inc., New York, 1958. The interesting problem of molecular weight distribution has been taken up by R. J. Zeman and N. R. Amundson, *Chem. Eng. Sci.*, **20**, 331, 637 (1965).

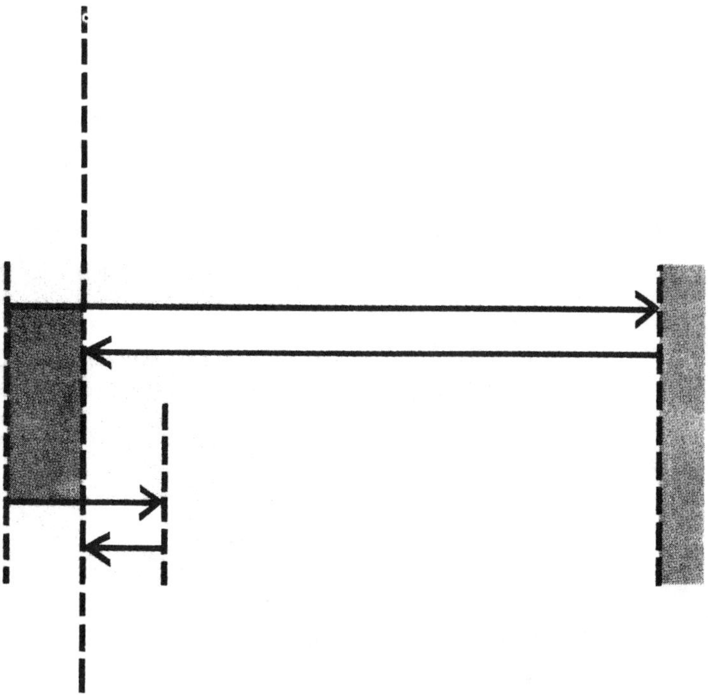

Simplified Kinetics of Sequences at the Steady State

4

4.1 *The Rate-determining Step*

Even with a small number of steps, the rate expression of a sequence is rather complicated in spite of the steady-state approximation. A simplification is in order whenever possible. Frequently, for a restricted range of experimental conditions, it can be assumed with considerable success that a given step in the sequence is rate-determining. If this occurs, all the other steps in the sequence will be in quasi-equilibrium and the kinetic problem is reduced to a consideration of the kinetics of a single step and the thermodynamic equilibrium of all the other steps. Sequences that are not amenable to the general treatment of the preceding chapter can still be treated readily.

Consider, for instance, the closed catalytic sequence:

$$\mathbf{S} + H_2O \rightleftarrows \mathbf{S{-}O} + H_2 \qquad (1)$$

$$\mathbf{S{-}O} + CO \rightleftarrows \mathbf{S} + CO_2 \qquad (2)$$

taking place at the surface of a solid catalyst with active centers S.
At the steady state:

$$r_1 = \vec{r}_1 - \overleftarrow{r}_1 = r_2 = \vec{r}_2 - \overleftarrow{r}_2$$

This situation might be realized as follows:

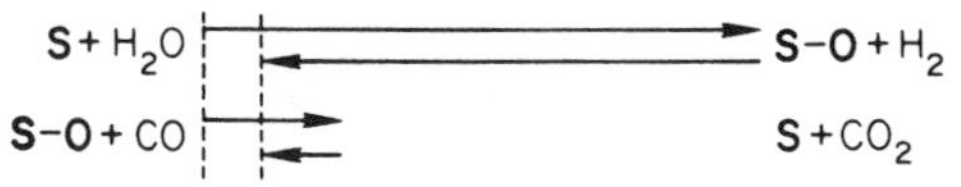

$$S + H_2O \rightleftharpoons S\text{-}O + H_2 \qquad (1)$$
$$S\text{-}O + CO \rightleftharpoons S + CO_2 \qquad (2)$$

where the length of each arrow is proportional to the corresponding rate.
While r_1 is equal to r_2 at the steady state both $\vec{r}_1$ and $\overleftarrow{r}_1$ are very much larger
than $\vec{r}_2$ and $\overleftarrow{r}_2$. Furthermore, reaction (1) is very near equilibrium; if it were
at equilibrium, $\vec{r}_1$ would be equal to $\overleftarrow{r}_1$. In fact, the quantity

$$\frac{\vec{r}_1 - \overleftarrow{r}_1}{\vec{r}_1} = \frac{r}{\vec{r}_1} \qquad (4.1.1)$$

a measure of the distance of reaction (1) from equilibrium, is much smaller
than the corresponding quantity for reaction (2):

$$\frac{\vec{r}_2 - \overleftarrow{r}_2}{\vec{r}_2} = \frac{r}{\vec{r}_2} \qquad (4.1.2)$$

since $\vec{r}_1 \gg \vec{r}_2$. A concise description of the situation is contained in the two
complementary statements: step (1) is in quasi-equilibrium; step (2) is rate-
determining.

With this additional assumption, we can write:

$$(S)a_1 = (S\text{-}O)a_{-1}$$

or since $(S) + (SO) = (L)$, the total concentration of sites:

$$(S\text{-}O) = (L) \frac{\left(\dfrac{a_1}{a_{-1}}\right)}{1 + \left(\dfrac{a_1}{a_{-1}}\right)} \qquad \text{and} \qquad (S) = (L) \frac{1}{1 + \left(\dfrac{a_1}{a_{-1}}\right)} \qquad (4.1.3)$$

The ratio $(S\text{-}O)/(L)$ is the fraction of surface sites covered with oxygen
atoms. This fraction is usually denoted by θ, with a subscript, if necessary,
denoting the nature of the species involved.

The rate expression then becomes:

$$r = r_2 = (\mathrm{L})a_2 \frac{\left(\frac{a_1}{a_{-1}}\right)}{1 + \left(\frac{a_1}{a_{-1}}\right)} - (\mathrm{L})a_{-2} \frac{1}{1 + \left(\frac{a_1}{a_{-1}}\right)} \tag{4.1.4}$$

or, more compactly:

$$r = (\mathrm{L}) \frac{a_1 a_2 - a_{-1} a_{-2}}{a_1 + a_{-1}} \tag{4.1.5}$$

It can be seen that, if there is a rate-determining step, only the rate constants of that step enter into the rate equation. The rate constants of the other steps in quasi-equilibrium appear only as ratios which, as is already known, are equal to the equilibrium constants of these equilibrated steps. This is a considerable simplification of the kinetic problem, since, at least in principle, equilibrium constants are more easily arrived at than rate constants. Even when this is not so, the number of arbitrary constants in the rate equation is reduced considerably.

The advantage of the simplifying assumption becomes even more obvious in the case of closed sequences where the elementary steps are of order higher than one with respect to the active centers.

Consider for instance the closed sequence:

$$\begin{aligned} 2\,\mathbf{S} + O_2 &\rightleftarrows 2\,\mathbf{SO} \qquad (1) \\ \mathbf{SO} + SO_2 &\rightleftarrows SO_3 + \mathbf{S} \qquad (2) \end{aligned} \tag{4.1.6}$$

taking place at the surface of a solid catalyst. The second step must be taken twice in order to reproduce, by summation of both steps, the equation for the stoichiometrically simple reaction:

$$2\,SO_2 + O_2 \rightleftarrows 2\,SO_3$$

taking place through the active centers **S** and **SO**.

From the definition of rates:

$$r_1 = \tfrac{1}{2}\frac{d(\mathrm{SO})}{dt} \qquad \text{and} \qquad r_2 = -\frac{d(\mathrm{SO})}{dt}$$

Thus, at the steady state,

$$r_1 = \tfrac{1}{2}\, r_2$$

More explicitly,

$$a_1(\mathrm{S})^2 - a_{-1}(\mathrm{SO})^2 = \tfrac{1}{2}\, a_2(\mathrm{SO}) - \tfrac{1}{2}\, a_{-2}(\mathrm{S})$$

This quadratic equation, together with (S) + (SO) = (L), leads to a rate equation which does not have the simple form of the previous solutions dealing with sequences where all steps are first-order with respect to active centers.

On the other hand, if it is assumed that step (1) is rate-determining and therefore that step (2) is in quasi-equilibrium, the latter condition gives:

$$a_2(\mathrm{SO}) = a_{-2}(\mathrm{S})$$

This, together with the usual relation (S) + (SO) = (L) gives:

$$(\mathrm{S}) = (\mathrm{L})\,\frac{1}{1 + \left(\dfrac{a_{-2}}{a_2}\right)} \qquad (\mathrm{SO}) = (\mathrm{L})\,\frac{\left(\dfrac{a_{-2}}{a_2}\right)}{1 + \left(\dfrac{a_{-2}}{a_2}\right)}$$

Consequently:

$$r = r_1 = a_1(\mathrm{S})^2 - a_{-1}(\mathrm{SO})^2$$

$$= (\mathrm{L})^2\left\{a_1\left[\frac{1}{1 + \left(\dfrac{a_{-2}}{a_2}\right)}\right]^2 - a_{-1}\left[\frac{\left(\dfrac{a_{-2}}{a_2}\right)}{1 + \left(\dfrac{a_{-2}}{a_2}\right)}\right]^2\right\} \tag{4.1.7}$$

This result indicates that the rate should be proportional to the square of the concentration of active centers at the surface. But this is due to the literal application of rate theory to step (1). In fact, an oxygen molecule does not require any two sites to be adsorbed; these sites have to be nearest neighbors. If each site is surrounded by Z sites, as determined by the structure of the solid, the rate of adsorption of oxygen, r, should be proportional to: (a) the concentration of oxygen, (b) the concentration c^f free sites (S), and (c) to the probability that any of the surrounding sites is free, i.e., $\frac{1}{2} Z(1 - \theta)$ where

$$1 - \theta = \frac{(\mathrm{S})}{(\mathrm{S}) + (\mathrm{SO})} = \frac{(\mathrm{S})}{(\mathrm{L})} \tag{4.1.8}$$

is the fraction of surface sites which is free. The factor $\frac{1}{2}$ is introduced because of the indistinguishability of neighboring sites and avoids counting sites twice.

We see that

$$r_1 = \tfrac{1}{2}\, Z k_1(\mathrm{O}_2)(\mathrm{S})\,\frac{(\mathrm{S})}{(\mathrm{L})} \tag{4.1.9}$$

Similarly, we have

$$r_{-1} = \tfrac{1}{2} \mathcal{Z} k_{-1}(\mathrm{SO}) \frac{(\mathrm{SO})}{(\mathrm{L})} \tag{4.1.10}$$

Therefore $(\mathrm{L})^2$ in (4.1.7) should be replaced by $\frac{1}{2} \mathcal{Z}(\mathrm{L})$. In practice, however, with solids, it is generally impossible to vary (L) which is a constant for a given solid catalyst. It is convenient to put $(\mathrm{L}) = 1$ and forget about this factor entirely since it is not a variable that can be controlled. This simplification, which will be made from now on, does not imply any loss of generality. The unknown value of $\frac{1}{2} \mathcal{Z}(\mathrm{L})$ is simply absorbed in the rate constants, in this case a_1 and a_{-1} and would be needed only in quantitative estimates.

It is interesting to note that the rate expression (4.1.7) is still of the general form that was derived in the preceding chapter. Indeed, it can be rewritten as:

$$r = \frac{a_1 a_2^2 - a_{-1} a_{-2}^2}{(a_2 + a_{-2})^2} \tag{4.1.11}$$

The numerator, as before, is an expression of the form of the law of mass action and the denominator is the square of a sum of terms involving concentrations of reactants and products. This is seen more clearly by expressing the a's in terms of rate constants and concentrations:

$$r = \frac{k_1 k_2^2 (\mathrm{O_2})(\mathrm{SO_2})^2 - k_{-1} k_{-2}^2 (\mathrm{SO_3})^2}{[k_2(\mathrm{SO_2}) + k_{-2}(\mathrm{SO_3})]^2} \tag{4.1.12}$$

or

$$r = \frac{\vec{k}(\mathrm{O_2})(\mathrm{SO_2})^2 - \overleftarrow{k}(\mathrm{SO_3})^2}{[k_2(\mathrm{SO_2}) + k_{-2}(\mathrm{SO_3})]^2} \tag{4.1.13}$$

with $\vec{k} = k_1 k_2^2$ and $\overleftarrow{k} = k_{-1} k_{-2}^2$. (4.1.14)

In particular, at equilibrium, $r = 0$, and

$$\frac{(\mathrm{SO_3})_e^2}{(\mathrm{O_2})_e (\mathrm{SO_2})_e^2} = K_c = \frac{\vec{k}}{\overleftarrow{k}} \tag{4.1.15}$$

Again, the ratio of rate constants for the forward and backward reactions is found to be equal to the equilibrium constant for the reaction, as was the case for an elementary step. This conclusion, which is not generally correct, will be taken up again in the next section.

Problem 4.1.1

Comparison of Eq. (4.1.5) with the general expression derived without the assumption of a rate-determining step reveals that the latter reduces to its simplified form if a_2 and a_{-2} are neglected in the denominator. Show that this follows naturally from the condition used in the definition of the rate-determining step: namely that both $\vec{r}_1$ and $\overleftarrow{r}_1$ are much larger than $\vec{r}_2$ and $\overleftarrow{r}_2$.

Problem 4.1.2

Write down the rate equation for the catalytic oxidation of sulfur dioxide going through the closed sequence (4.1.6), on the assumption that the second step is rate-determining.

Problem 4.1.3

Using the result of Problem 4.1.2, find for that particular case, the relation between rate constants and equilibrium constant.

4.2 The Stoichiometric Number of the Rate-determining Step

If in a closed sequence it is possible under a given set of conditions to recognize a rate-determining step (subscript s), all the other steps being in quasi-equilibrium, the free-energy difference for the overall reaction ΔG will be proportional to the free-energy difference for the rate-determining step ΔG_s.

The factor of proportionality is equal to the *stoichiometric number of the rate-determining step:*

$$\Delta G = s\Delta G_s$$

This number s is the number of times that the rate-determining step, as written, must be repeated, in the closed sequence, in order to obtain by summation of all steps the overall stoichiometric equation for reaction, as written. It will be remembered that, according to the convention adopted in Section 2.1, there is no freedom to write the equation for elementary steps at will; it must be written in the way that the step takes place at the molecular level. This restriction does not exist regarding the overall equation for reaction which can always be written in an arbitrary way. As a consequence, the value of stoichiometric number of the rate-determining step is dictated by the way the stoichiometric equation for reaction has been written.

Now, sufficiently near equilibrium, for any elementary step, and in particular for the rate-determining step, it is always permitted to write according to (2.6.4):

$$r_s = \frac{(r_s)_e}{RT}(-\Delta G_s) \tag{4.2.1}$$

Since, however,

$$r = \frac{1}{s} r_s, \qquad r_e = \frac{1}{s}(r_e)_s \quad \text{and} \quad \Delta G = s\Delta G_s,$$

(4.2.1) can be rewritten in terms of s and of quantities pertaining to the overall reaction:

$$r = \frac{r_e}{sRT}(-\Delta G) \tag{4.2.2}$$

Now, Eq. (2.6.12) for the first-order rate constant k of an elementary step in the vicinity of equilibrium:

$$k = r_e \sum_i \frac{\nu_i^2}{(c_i)_e} \tag{2.6.12}$$

was derived from the relation:

$$r = \frac{r_e}{RT}(-\Delta G) \tag{2.6.4}$$

Therefore, in the case of a closed sequence for which there exists a rate-determining step with stoichiometric number s, the first-order rate constant describing the advancement of the overall reaction toward equilibrium, in the vicinity of equilibrium, becomes:

$$\boxed{k = \frac{r_e}{s} \sum_i \frac{\nu_i^2}{(c_i)_e}} \tag{4.2.3}$$

A study of the kinetics of the reaction near equilibrium can then lead to a determination of s if only the rate at equilibrium r_e can be measured independently. This can always be done in principle by studying the rate of exchange of a tracer atom between reactants and products of the system in chemical equilibrium. Thus the rate of the ammonia reaction at equilibrium:

$$N_2 + 3\,H_2 \leftrightarrow 2\,NH_3$$

can be found by a study of the kinetics of the isotopic exchange reaction:

$$N^{14}N^{14} + N^{15}H_3 \rightleftarrows N^{14}N^{15} + N^{14}H_3$$

taking place in a system containing nitrogen, hydrogen and ammonia in equilibrium concentrations.

In general, consider the exchange reaction:

$$AM + BM^* \rightleftarrows AM^* + BM$$

where M^* is an isotope of M, taking place in a system in a state of chemical equilibrium containing a constant number n_1 of molecules of the AM type and a constant number n_2 of molecules of the BM type. Let us denote by R_e the rate of the exchange reaction at equilibrium. It is assumed that, at zero time, there exist in the system $n^0_{BM^*}$ molecules of the BM^* type, no molecule of the n_{AM^*} type, n^0_{AM} and n^0_{BM} molecules of the AM and BM type respectively. Clearly:

$$n^0_{AM} = n_1$$

$$n^0_{BM} + n^0_{BM^*} = n_2$$

The extent of the exchange reaction, $X = n_{AM^*}$ will change at a net rate dX/dt proportional to R_e:

$$\frac{dX}{dt} = R_e \left[\frac{n_{AM}}{(n_{AM} + n_{AM^*})} \frac{n_{BM^*}}{(n_{BM} + n_{BM^*})} - \frac{n_{AM^*}}{(n_{AM} + n_{AM^*})} \frac{n_{BM}}{(n_{BM} + n_{BM^*})} \right] \tag{4.2.4}$$

The factor of proportionality consists of two terms, corresponding to the exchange from left to right and from right to left. The rate of exchange from left to right is proportional to the fraction of molecules of the AM type that are in the AM^* form and to the fraction of molecules of the BM type that are in the BM^* form. Similarly for the reverse reaction. This argument is based on a straightforward consideration of probability and does not depend on any assumed sequence of the steps required for the exchange.

At any time t:

$$n_{AM} + n_{AM^*} = n_1$$

$$n_{BM} + n_{BM^*} = n_2$$

$$n_{AM} = n^0_{AM} - X$$

$$n_{BM^*} = n^0_{BM^*} - X$$

$$n_{AM^*} = X$$

$$n_{BM} = n^0_{BM} + X$$

The expression (4.2.4) then becomes:

$$\frac{dX}{dt} = \frac{R_e}{n_1 n_2}\left[(n^0_{AM} n^0_{BM^*}) - X(n_1 + n_2)\right] \tag{4.2.5}$$

But $n^0_{AM} = n_1$ and at sufficiently large values of time, $t = \infty$, exchange is complete so that:

$$\frac{n^\infty_{AM^*}}{n^\infty_{BM^*}} = \frac{n_1}{n_2}$$

or

$$\frac{n^\infty_{AM^*}}{n^\infty_{AM^*} + n^\infty_{BM^*}} = \frac{n_1}{n_1 + n_2}$$

Also:

$$n^\infty_{AM^*} + n^\infty_{BM^*} = n^0_{BM^*}$$

$$n^\infty_{AM^*} = X_\infty$$

Therefore:

$$\frac{dX}{dt} = R_e \frac{n_1 + n_2}{n_1 n_2}(X_\infty - X) \tag{4.2.6}$$

After integration:

$$\boxed{\ln \frac{X_\infty}{X_\infty - X} = R_e \frac{n_1 + n_2}{n_1 n_2} t} \tag{4.2.7}$$

Thus measurements of X as a function of time will yield the value of R_e. In principle, it is possible to determine the stoichiometric number of the rate-determining step by measurements of the rate near and at equilibrium. In practice, the nature of the rate-determining step and therefore the value of s are unknown. They must be postulated. As a result of the assumptions made, an expression for the rate can be derived. This expression is confronted with experimental data; if the fit is satisfactory, a presumption exists that the postulated sequence and, in particular, the postulated rate-determining step may be at least partially correct.

Frequently, the procedure is even more empirical, especially at the start of a kinetic investigation. It is simply postulated that the rate will be represented within a limited range of conversion, pressure or composition by an expression of the type of the mass action law, similar to (1.4.3) and (1.4.5):

$$r = \vec{r} - \overleftarrow{r} = \vec{k} \prod_i c_i^{\vec{\alpha}_i} - \overleftarrow{k} \prod_i c_i^{\overleftarrow{\alpha}_i} \tag{4.2.8}$$

where the parameters $\vec{\alpha}_i$ and $\overleftarrow{\alpha}_i$ are integers or fractions, positive, negative, or zero, for all components of the system and are determined from available kinetic data (Rule IV).

Justification for the approximate validity can be found in the results obtained in this chapter and the preceding one. For general sequences, the rate can be expressed as a ratio; the numerator is an expression of the form of the law of mass action and the denominator is a sum (which may be squared) of terms, each term being also of the form of the law of mass action. It is natural to expect, and it is corroborated by experience, that this ratio can be approximately expressed by a simple expression of the form of the mass action law with exponents that have no physical meaning but are simply adjustable parameters.

Can the parameters $\vec{\alpha}_i$ and $\overleftarrow{\alpha}_i$ take any arbitrary value? The answer is no; there exists a relation between them. Indeed, at equilibrium, $r = 0$ and

$$\frac{\vec{k}}{\overleftarrow{k}} = \prod_i \frac{c_i^{\overleftarrow{\alpha}_i}}{c_i^{\vec{\alpha}_i}}$$

But $\vec{k}/\overleftarrow{k}$ depends only on temperature. The equilibrium constant also depends only on temperature:

$$K_c = \prod_i c_i^{\nu_i}$$

Consequently $\prod_i c_i^{\overleftarrow{\alpha}_i}/c_i^{\vec{\alpha}_i}$ must be a certain function f of $\prod_i c_i^{\nu_i}$:

$$\prod_i \frac{c_i^{\overleftarrow{\alpha}_i}}{c_i^{\vec{\alpha}_i}} = f\left(\prod_i c_i^{\nu_i}\right)$$

This relation must be valid at all values of the concentrations. This will be so if:

$$\prod_i \frac{c_i^{\overleftarrow{\alpha}_i}}{c_i^{\vec{\alpha}_i}} = \left(\prod_i c_i^{\nu_i}\right)^n$$

with $\overleftarrow{\alpha}_i - \vec{\alpha}_i = \nu_i n$ for all values of i. In particular, then, there exists a relation between rate constants and equilibrium constants of the form

$$\boxed{\frac{\vec{k}}{\overleftarrow{k}} = K_c^{\,n}} \tag{1.4.5}$$

This relation was called Rule V in the first chapter.

Let us further inquire into any restriction as to the value of n. Because of the last equation, following now familiar arguments, it is easy to show that:

$$r = \vec{r}\left[1 - \exp\left(n\frac{\Delta G}{RT}\right)\right] \tag{4.2.9}$$

which near equilibrium reduces to:

$$r = \frac{r_e n}{RT}(-\Delta G) \tag{4.2.10}$$

Clearly then, n must be positive, otherwise the rate would be positive for positive changes in free energy, which is contrary to thermodynamics: If ΔG is positive, a reaction is impossible. Thus n must be a positive integer or fraction.

If besides there exists an identifiable rate-determining step with stoichiometric number s, the value of n has a definite physical meaning. Indeed, near equilibrium, as was just seen:

$$r = \frac{r_e}{sRT}(-\Delta G) \tag{4.2.2}$$

Therefore n must be equal to $1/s$. Also:

$$\boxed{\frac{\vec{k}}{\overleftarrow{k}} = K^{1/s}} \tag{4.2.11}$$

This is a very useful relation because it restricts considerably the choice of adjustable parameters in a rate expression of the form of the law of mass action.

Problem 4.2.1

If the catalytic oxidation of SO_2 takes place through the sequence assumed in Problems 4.1.2 and 4.1.3, what is the stoichiometric number of the rate-determining step in both cases, if the stoichiometric equation for the overall reaction is written as:

$$SO_2 + \tfrac{1}{2}O_2 \rightleftarrows SO_3$$

Problem 4.2.2

Find an expression for s, the stoichiometric number of the rate-determining step of the reaction

$$N_2 + 3H_2 \rightleftarrows 2NH_3$$

in terms of k, the measurable first-order rate constant describing the kinetics of the reaction near equilibrium, and of X, X_∞ and t as they enter into the equation just derived, measured for the exchange reaction

$$N^{14}N^{14} + N^{15}H_3 \rightleftarrows N^{15}N^{14} + N^{14}H_3$$

taking place in a system containing equilibrium concentrations of N_2, H_2 and NH_3. Note that if the ammonia reaction is studied at say 1 atm and 450°C, the equilibrium amount of ammonia in the system is extremely small. Find the expression for s corresponding to this simplified situation.

Problem 4.2.3

Show that for the reaction

$$H_2 + Br_2 \rightarrow 2\ HBr$$

the exact rate expression derived in Problem 3.4.2 can be put in the form:

$$r = k(H_2)^{\alpha_1}(Br_2)^{\alpha_2}(HBr)^{\alpha_3}$$

where $\alpha_1 = 1$, $\alpha_2 = \frac{3}{2}$, $\alpha_3 = -1$. Under what conditions would this approximation be satisfactory?

Problem 4.2.4

It has been found that, under certain conditions over a certain catalyst, the forward rate of catalytic oxidation of SO_2 can be represented by the expression

$$\vec{r} = \vec{k}(O_2)(SO_2)^{1/2}(SO_3)^{-1/2}$$

Find an expression for $\overleftarrow{r}$ assuming only that reaction of O_2 with active centers at the surface is rate-determining.

4.3 *Shift in Rate-determining Step*

While the steady-state approximation is rarely in error and constitutes an almost universal tool for the kinetic treatment of sequences involving active centers, the assumption of a rate-determining step must be made with great caution. If such a step exists at all, there is no guarantee that it will not shift as process conditions are changed. The variables affecting rates are temperature, pressure and composition, and hence a shift in rate-determining step may occur as these variables are changed.

Consider a simple example where *temperature* will cause a shift in rate-determining step. When hydrogen molecules, at low pressures, collide with a hot metallic surface (for example, tungsten), hydrogen atoms are produced catalytically by the two-step sequence:

$$H_2 + 2\,\mathbf{S} \rightarrow 2\,\mathbf{SH} \qquad (1)$$

$$\mathbf{SH} \rightarrow \mathbf{S} + H \qquad (2)$$

If step (2) is taken twice, the overall reaction is:

$$H_2 \rightarrow 2\,H$$

At high temperatures where the rate of the reaction is still slow but measurable, the rate-determining step is (2); the first step is in quasi-equilibrium and thus:

$$a_1(\mathrm{S})^2 = a_{-1}(\mathrm{SH})^2$$

But because the temperature is already high, the concentration (SH) is very small. Indeed step (1) is exothermic from left to right so that equilibrium is shifted more and more to the left as the temperature increases. Therefore, it is a good approximation to write $(\mathrm{S}) \cong (\mathrm{L}) = (1)$ and:

$$(\mathrm{SH}) = \left(\frac{a_1}{a_{-1}}\right)^{1/2}$$

Then

$$r = \tfrac{1}{2}\,r_2 = \tfrac{1}{2}\,a_2(\mathrm{SH}) = \tfrac{1}{2}(\mathrm{L})a_2\left(\frac{a_1}{a_{-1}}\right)^{1/2}$$

This can be rewritten as:

$$r = \tfrac{1}{2}\,k_2\left(\frac{k_1}{k_{-1}}\right)^{1/2}(H_2)^{1/2} \qquad (4.3.1)$$

The situation might be depicted as follows:

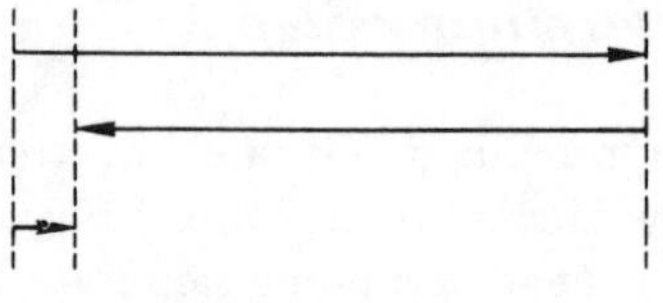

(1)

(2)

where the length of the arrows are proportional to rates.

At still higher temperatures, the situation becomes:

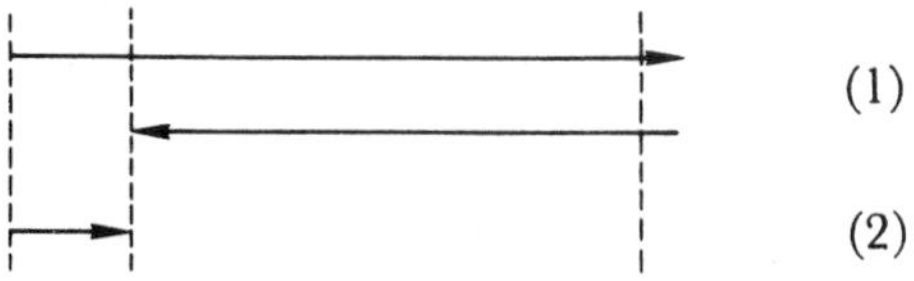

It is seen that $\vec{r}_2 = r_2$ has increased considerably more than $\vec{r}_1$ or $\overleftarrow{r}_1$ because it has a higher activation energy. The assumption that $\vec{r}_1$ and $\overleftarrow{r}_1$ are much larger than r_2 becomes poor. Finally, at very high temperatures, $\overleftarrow{r}_1$ becomes negligibly small because (SH) has now become vanishingly small:

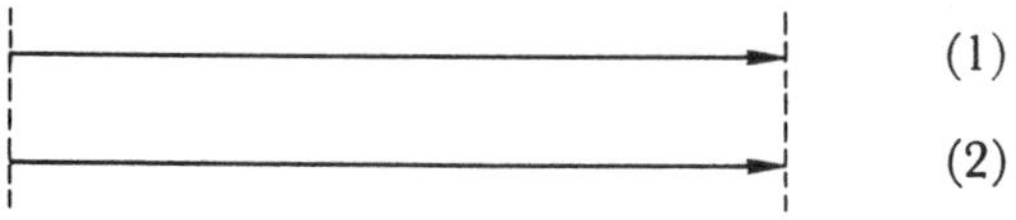

Consequently $r_1 = \vec{r}_1 = r_2 = r$:

$$r = a_1(\mathrm{S})^2 = a_1(\mathrm{L})^2 = k_1(\mathrm{H_2})$$

The rate expression has changed correspondingly since the assumption that step (2) is the rate-determining step has now become invalid.

A very similar illustration of the shift in rate-determining step with *pressure* deals with the important problem of the unimolecular transformation of a molecule A into one or several products P: A $\rightarrow$ P. As first proposed by Lindemann(1922), it proceeds by means of an open sequence of two steps:

$$\mathrm{A} + \mathrm{A} \rightleftarrows \mathrm{A}^* + \mathrm{A} \qquad (1)$$

$$\mathrm{A}^* \rightarrow \mathrm{P} \qquad (2)$$

In the first step, the energy required for reaction is accumulated into molecule A by means of collisions with other molecules A to yield an energized molecule A*, i.e., a molecule that has sufficient energy to react. This energized molecule then decomposes in the second step.

Applying the steady-state approximation to this open sequence where A* is an active center, we get:

$$r = \frac{a_1 a_2(\mathrm{A})}{a_{-1} + a_2} \qquad (4.3.2)$$

Equation (4.3.2) can be rewritten in the more explicit form:

$$r = \frac{k_1 k_2(\mathrm{A})^2}{k_{-1}(\mathrm{A}) + k_2} \qquad (4.3.3)$$

If the pressure is sufficiently high $k_{-1}(A) \gg k_2$ and

$$r = \left(\frac{k_1}{k_{-1}}\right) k_2(A) \tag{4.3.4}$$

The unimolecular reaction is also first-order. The first step is in quasi-equilibrium and, as a consequence, the fundamental assumption of equilibrium applies. Thus $(k_1/k_{-1})k_2 = k$ where k is given by transition-state theory.

The situation may again be represented schematically as follows:

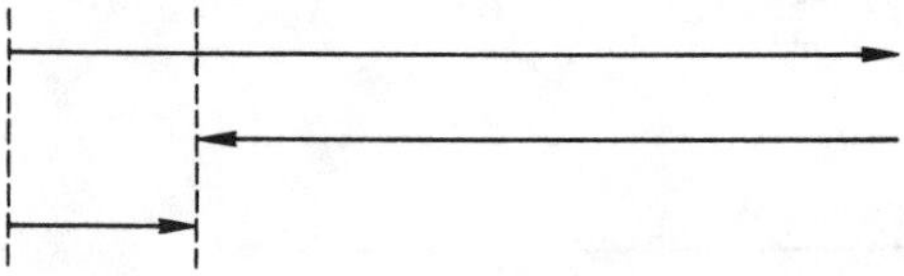

(1)

(2)

But as pressure is lowered, $\vec{r}_1$ and $\overleftarrow{r}_1$ that are proportional to the square of pressure decrease faster than r_2 which is simply proportional to pressure:

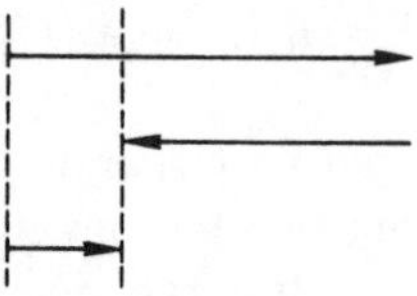

(1)

(2)

The assumption that (2) is rate-determining becomes poor. At sufficiently low pressure, the inequality $\vec{r}_1$ (or $\overleftarrow{r}_1$) $\gg r_2$ breaks down completely:

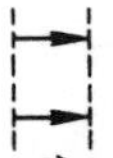

(1)

(2)

$r_1 = \vec{r}_1 = r_2$

Also then $k_{-1}(A) \ll k_2$ and

$$r = k_1(A)^2 \tag{4.3.5}$$

The unimolecular reaction has become second-order, in apparent contradiction of expectation. The reason is of course that A → P is not an elementary step. In particular, the fundamental equilibrium assumption breaks down and the rate is now that of an energy transfer process, beyond the scope of transition-state theory. Behavior at low pressure may be qualified as abnormal.

The pressure range at which transition occurs between normal and abnormal behavior depends on the relative magnitudes of $k_{-1}(A)$ and k_2. The value of k_2 decreases as molecular size increases because it takes more time for the energy, once it has been accumulated by collisions, to fluctuate into the proper reaction coordinate to bring about reaction. Thus for molecules containing four or five atoms, the transition will lie at relatively low pressures, normally sizeably lower than those used in practical applications. On the other hand, if the molecule is diatomic, the value of k_2 is very large because the molecule will react as soon as it receives the necessary energy. Consequently, for diatomic molecules, unimolecular reaction is always second-order at all practical pressures. Reaction is here decomposition into atoms. In fact, collision with the molecules A_2 themselves or with any other molecule M in the system will serve to energize A_2, so that the decomposition may be represented as follows:

$$A_2 + M \rightarrow A + A + M$$
$$r = k(A_2)(M)$$

An important consequence of this situation is that recombination of atoms to form a diatomic molecule necessitates a molecule M, called third body, to remove the energy of bond formation. At equilibrium, this results naturally from the situation just discussed for decomposition by virtue of the principle of microscopic reversibility:

$$A + A + M \rightarrow A_2 + M \tag{4.3.6}$$
$$r = k(A)(A)(M)$$

The recombination of atoms is therefore a third-order process. An order of magnitude for k is

$$k = 10^{-32}\ \text{cm}^6/\text{sec}$$

This value can be understood by a simple estimation. At atmospheric pressure, the mean free path in a gas is about 10^{-5}cm while $(M) \cong 10^{19}\text{cm}^{-3}$. The probability of a triple collision is that of a binary collision times a factor approximately equal to the ratio of molecular dimension to mean free path: $10^{-8}/10^{-5} = 10^{-3}$. Since the rate of binary collisions is ca. $10^{-10}(A)(A)$ cm^{-3} sec^{-1} (Table 2.5.1), the value of k will be $10^{-10} \times 10^{-3} \times 10^{-19} \cong 10^{-32}$ cm^6/sec.

For recombination of polyatomic species, there is no need for a third body, just as decomposition of a polyatomic molecule is a first-order reaction at all reasonable pressures. As an example, the recombination of methyl radicals:

$$CH_3 + CH_3 \rightarrow C_2H_6$$

is described by the second-order rate expression:

$$r = k(CH_3)^2$$

Finally, *composition*, as for instance expressed by the extent of reaction, may cause a shift in the rate-determining step. Consider again the propagation steps in the hydrogen-bromine system:

$$\mathbf{Br} + H_2 \rightleftarrows HBr + \mathbf{H} \qquad (1)$$

$$\mathbf{H} + Br_2 \rightarrow HBr + \mathbf{Br} \qquad (2)$$

It is known that $k_2 = 10\,k_{-1}$ over a large range of temperatures. If the reaction takes place in a system containing reactants and products in the ratio

$$(H_2) : (Br_2) : (HBr) = 1 : 1 : 200$$

as would be the case very near equilibrium in a system containing originally equimolar amounts of H_2 and Br_2, and if r_2 is taken to be unity for comparison, then $\overleftarrow{r}_1 = 20\,r_2$. But at steady state: $\overrightarrow{r}_1 - \overleftarrow{r}_1 = r_1 = r_2$. Therefore $\overrightarrow{r}_1 = 21\,r_2$ and we can represent the situation schematically as:

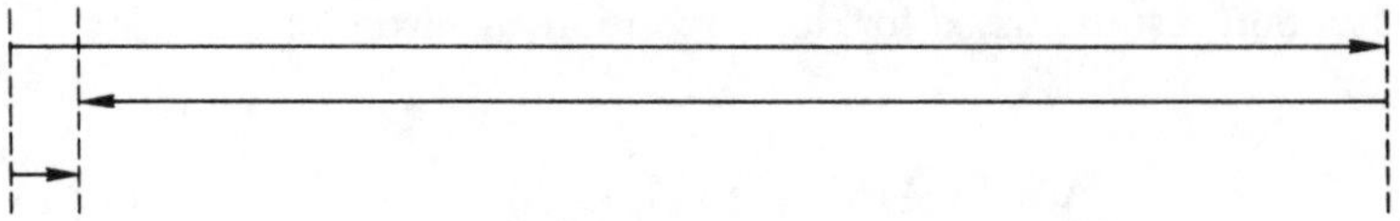

(1)

(2)

Clearly, it can be said that step (2) is rate-determining and step (1) is in quasi-equilibrium. But again, if the system is examined at smaller and smaller values of conversion, the assumption gets poorer and poorer. In the initial system, with reactants only, the situation could be pictured as follows:

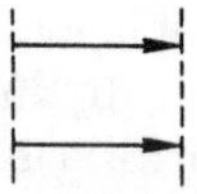

(1)

(2)

There is no rate-determining step. A rate-determining step will only appear at sufficiently high degrees of conversion.

In summary, the assignment of a specific step as rate-determining presents such a simplification in the kinetic treatment of sequences, that this approximation will be attempted whenever possible. But it must be kept in mind that the nature of the rate-determining step may well change with reaction conditions, and careless extrapolation can lead to serious error.

Problem 4.3.1

Verify (4.3.2).

Problem 4.3.2

Find the pressure dependence of the rate of the reaction

$$H_2 + Br_2 \rightarrow 2\ HBr$$

if the active centers (bromine atoms) are produced at a rate

$$r_i = I(Br_2)$$

by means of light absorbed by bromine molecules. The constant I does not depend on pressure. Termination is by three-body recombination of bromine atoms.

4.4 *Rate-determining Active Centers*

According to the concept of a rate-determining step, if there is one, all other steps in a sequence are in quasi-equilibrium. Thus these other steps must be reversible. Consider now a closed sequence where all steps are irreversible:

$$A_1 + \mathbf{X}_1 \rightarrow \mathbf{X}_2 + B_1 \qquad (1)$$

$$A_2 + \mathbf{X}_2 \rightarrow \mathbf{X}_3 + B_2 \qquad (2)$$

$$A_3 + \mathbf{X}_3 \rightarrow \mathbf{X}_4 + B_3 \qquad (3)$$

$$A_4 + \mathbf{X}_4 \rightarrow \mathbf{X}_1 + B_4 \qquad (4)$$

There can be no question in this case of a rate-determining step in the sense used in this chapter. Yet there may exist a rate-determining active center, say X_1. This will be the one present at highest concentration, so that $(X_1) \gg (X_2,)\ (X_3,)\ (X_4.)$ It follows directly from (3.3.4) that:

$$\frac{(X_1)}{(X_2)} = \frac{M_1}{M_2} \qquad \frac{(X_1)}{(X_3)} = \frac{M_1}{M_3} \qquad \frac{(X_1)}{(X_4)} = \frac{M_1}{M_4} \qquad (4.4.1)$$

For irreversible steps, the values of M_1, M_2, M_3, M_4 are simply $a_2a_3a_4$, $a_3a_4a_1$, $a_4a_1a_2$, $a_1a_2a_3$ so that:

$$\frac{(X_1)}{(X_2)} = \frac{a_2}{a_1} \qquad \frac{(X_1)}{(X_3)} = \frac{a_3}{a_1} \qquad \frac{(X_1)}{(X_4)} = \frac{a_4}{a_1} \qquad (4.4.2)$$

Therefore, the condition $(X_1) \gg (X_2,)\,(X_3,)\,(X_4)$ will be satisfied if $a_1 \ll a_2, a_3, a_4$. Then X_1, the least reactive active center, will be also the most abundant.

It is very useful to know whether this situation prevails, because such knowledge will greatly facilitate the kinetic treatment of chain reactions. Indeed, for chain reactions with long chains, the steady-state concentration of active centers is determined by the relation $r_i = r_t$ where r_i represents the rate of initiation and r_t the rate of termination. Termination includes the destruction of all types of active centers present in the system. Clearly, the kinetic treatment will be simplified if only one termination step needs to be taken into account because it is the one that expresses the rate of destruction of the most abundant active centers.

Consider, for instance, the reaction between hydrogen and bromine at pressure such that termination takes place through recombination of the active centers in the gas phase, pairwise. There are two types of active centers, hydrogen atoms and bromine atoms (see Problem 3.4.2) and therefore there are three possible termination steps, all involving a third body M as discussed in Section 4.3:

$$\mathrm{H + H + M \rightarrow H_2 + M} \qquad (1)$$

$$\mathrm{Br + Br + M \rightarrow Br_2 + M} \qquad (2)$$

$$\mathrm{H + Br + M \rightarrow HBr + M} \qquad (3)$$

If all three termination steps had to be taken into account, the rate expression would not be particularly simple. But in this system, at 300°C and with equimolar amounts of H_2 and Br_2, we can write at the beginning of reaction when both steps of the sequence are irreversible:

$$\frac{(\mathrm{Br})}{(\mathrm{H})} = \frac{a_2}{a_1} = \frac{k_2}{k_1}\frac{(\mathrm{Br_2})}{(\mathrm{H_2})} = \frac{k_2}{k_1}$$

The rate constants at 300°C are such that $(k_2/k_1) \cong 10^6$. Consequently, the rate of recombination $r_{t,1}$ will be 10^{12} less rapid and the rate $r_{t,3}$ will be 10^6 times less rapid than the rate of recombination of bromine atoms. We can write:

$$r_t = r_{r,1} + r_{t,2} + r_{t,3} = r_{t,2}$$

and this is indeed what has been assumed earlier in Problem 3.4.2.

There is another aspect of the question that must be stressed at this stage. Even if we were able to estimate k_2 and k_1 only within *two orders of magnitude* by using a rough version of transition-state theory (Chapter 2) and some of the correlations to be presented later (Chapter 8), the great simplification made above and based on an actual knowledge of rate constants (a most unusual situation) could still have been carried out. This illustrates a very important aphorism of applied chemical kinetics: *Calculations of rates are attempted not to be kept but to be rejected.* Thus terribly poor estimates prove to be very valuable.

If a given active center is known to be rate-determining under a certain set of conditions, there is, of course, again no guarantee that it will remain so under a different set of reaction conditions determined by pressure, temperature or composition.

A particularly striking example of this possible shift in the nature of the rate-determining active center is known in the case of the early stages of oxidation of hydrocarbons RH, in particular olefins, in the liquid phase at low temperatures.

The initiation of the chain is due to an unspecified process that yields alkyl free radicals R at a rate r_i. The closed sequence is then (see Table 3.1.1):

$$\mathbf{R} + O_2 \rightarrow \mathbf{ROO} \qquad (1)$$

$$\mathbf{ROO} + RH \rightarrow ROOH + \mathbf{R} \qquad (2)$$

There are three possible termination steps:

$$\mathbf{R} + \mathbf{R} \rightarrow \qquad (t1)$$

$$\mathbf{R} + \mathbf{ROO} \rightarrow \qquad (t2)$$

$$\mathbf{ROO} + \mathbf{ROO} \rightarrow \qquad (t3)$$

Products of these steps are immaterial: They are unreactive and therefore the steps destroy active centers. In the long chain approximation, the products of chain termination also represent only a very small fraction of total products.

Rate constants k_1 and k_2 for the propagation steps (1) and (2) are such that, at temperatures of interest:

$$k_1(O_2) = a_1 > k_2(RH) = a_2$$

at sufficiently high pressures of oxygen, but:

$$k_1(O_2) = a_1 < k_2(RH) = a_2$$

at sufficiently low pressures of oxygen. Two limiting cases can be recognized. At high oxygen pressures, the rate-determining active center is ROO and the significant termination step is (t3). At low oxygen pressures, the rate-determining active center is R and the significant termination step is (t1). In practice, even if pressure is high, there may be regions in the liquid that are oxygen-starved if mixing and agitation are not adequate. Therefore, mass transfer can have a substantial effect on the pertinent kinetics of the process. These effects will be discussed in Chapter 7.

Problem 4.4.1

Assume a rate of initiation r_i for the long-chain liquid-phase oxidation of cumene to cumyl hydroperoxide. Derive rate functions for the two limiting cases of low and high pressure of oxygen.

4.5 *Ambiguity of Simplified Kinetics*

Very frequently, the same simplified rate expression can be obtained by making the assumption of either a rate-determining step or of a rate-determining active center. Which assumption is the correct one must then be decided by examination of the physical meaning of the parameters or by independent evidence.

Consider a simple example, an isomerization A → B taking place at the surface of a solid catalyst. The postulated sequence is:

$$\begin{array}{lcll} \mathrm{S} + \mathrm{A} & \rightleftarrows & \mathrm{A{-}S} & (1) \\ \mathrm{A{-}S} & \rightleftarrows & \mathrm{B{-}S} & (2) \\ \mathrm{B{-}S} & \rightleftarrows & \mathrm{B} + \mathrm{S} & (3) \end{array} \tag{4.5.1}$$

The complete rate expression, following (3.3.9) is rather hopeless:

$$r = \frac{a_1a_2a_3}{a_2a_3 + a_3a_1 + a_1a_2 + a_{-1}a_3 + a_{-2}a_1 + a_{-3}a_2 + a_{-1}a_{-2} + a_{-2}a_{-3} + \alpha_{-3}a_{-1}} \tag{4.5.2}$$

A first simplifying assumption is to postulate that step (2) is rate-determining. Then the sequence (4.5.1) becomes:

$$\begin{array}{lcll} \mathrm{S} + \mathrm{A} & \leftrightarrow & \mathrm{A{-}S} & (1) \\ \mathrm{A{-}S} & \rightarrow & \mathrm{B{-}S} & (2) \\ \mathrm{B{-}S} & \leftrightarrow & \mathrm{B} + \mathrm{S} & (3) \end{array} \tag{4.5.3}$$

and all terms in the denominator of (4.5.2) containing a_2 or a_{-2} can be neglected. Then (4.5.2) simplifies to:

$$r = \frac{a_1a_2a_3}{a_3a_1 + a_{-1}a_3 + a_{-3}a_{-1}} \tag{4.5.4}$$

Dividing both numerator and denominator of (4.5.4) by $a_{-1}a_3$, we get:

$$r = \frac{\dfrac{a_1}{a_{-1}} a_2}{1 + \dfrac{a_1}{a_{-1}} + \dfrac{a_{-3}}{a_3}} \tag{4.5.5}$$

We further assume that the surface concentration of **A—S** is much larger than that of **B—S** so that

$$\frac{a_1}{a_{-1}} \gg \frac{a_{-3}}{a_3}$$

Hence, Eq. (4.5.5) becomes:

$$r = \frac{\dfrac{a_1}{a_{-1}} a_2}{1 + \dfrac{a_1}{a_{-1}}} \tag{4.5.6}$$

which can be put in the form:

$$r = \frac{k(\mathrm{A})}{1 + k'(\mathrm{A})} \tag{4.5.7}$$

where

$$k = \frac{k_1}{k_{-1}} k_2 \qquad k' = \frac{k_1}{k_{-1}} \tag{4.5.8}$$

Assume now on the other hand that in the sequence (4.5.1) there is no rate-determining step but that all three steps are irreversible:

$$\begin{aligned} \mathbf{S} + \mathrm{A} &\rightarrow \mathbf{A}\text{—}\mathbf{S} \\ \mathbf{A}\text{—}\mathbf{S} &\rightarrow \mathbf{B}\text{—}\mathbf{S} \\ \mathbf{B}\text{—}\mathbf{S} &\rightarrow \mathrm{B} + \mathbf{S} \end{aligned} \tag{4.5.9}$$

Then all terms in the denominator of (4.5.2) that contain a_{-1}, a_{-2} or a_{-3} can be neglected so that:

$$r = \frac{a_1 a_2 a_3}{a_2 a_3 + a_3 a_1 + a_1 a_2} \tag{4.5.10}$$

or

$$r = \frac{a_1}{1 + \dfrac{a_1}{a_2} + \dfrac{a_1}{a_3}} \tag{4.5.11}$$

If, furthermore, it is assumed that **B—S** is the rate-determining surface complex so that the two main active centers will be **S** and **B—S**, we have $a_3 \ll a_1 \ll a_2$ and (4.5.11) simplifies to:

$$r = \frac{a_1}{1 + \dfrac{a_1}{a_3}} \tag{4.5.12}$$

which can be put in the form

$$r = \frac{k(\mathrm{A})}{1 + k'(\mathrm{A})} \tag{4.5.13}$$

where

$$k = k_1 \qquad k' = \frac{k_1}{k_3} \tag{4.5.14}$$

Comparison of (4.5.7) and (4.5.13) shows that identical rate expressions are obtained from two radically different sets of simplifying assumptions. It may be possible however to decide between the two alternatives by further examination of the parameters k and k' which have different interpretations (4.5.8) and (4.5.14) in both cases.

The example just discussed illustrates the ambiguity of simplified rate expressions, a very general problem of the kinetic analysis of sequences. More specifically, it shows that in the case of reactions taking place at the surface of solid catalysts, sequences of the type (4.5.3) that postulate adsorption equilibrium between the surface and reactants or products, lead to rate equations (4.5.6) which can also be obtained from sequences of the type (4.5.9). But the latter do not assume any adsorption equilibrium. This ambiguity is frequently expressed by a warning that mere kinetics can contribute no proof of mechanism. In this case "mechanism" means the proper sequence of steps. This warning must always be kept in mind: Chemical kinetics is a necessary but insufficient tool in the study of chemical change.

In conclusion, before making any assumption as to the existence of a rate-determining step, it is much safer and far more logical to write down, whenever possible, a complete rate expression. This expression will then be simplified according to the basic ideas of this chapter. These rely not only on the possible existence of a rate-determining step but also on that of a rate-determining active center.

Problem 4.5.1

The reaction A → B takes place at the surface of a solid catalyst. It is found that the rate of reaction is zero-order. Propose a sequence which will account for this result under two different sets of simplifying assumptions.

BIBLIOGRAPHY

4.1 The author was first introduced to the graphical representation used in this chapter by Kenzi Tamaru.

4.2 The theorem expressed in (4.2.10) was first demonstrated by Horiuti who also elaborated the method of measurement of s. An account of his work is found in *J. Catalysis*, **1**, 199 (1962). The concept of the stoichiometric number of the rate-determining step has been used widely in the kinetic treatment of electrode processes: See, for instance, B. E. Conway, *Theory and Principles of Electrode Processes*, The Ronald Press Company, New York, 1965.

4.3 The problem of unimolecular reactions is an important and active chaper in pure chemical kinetics: S. W. Benson, *The Foundations of Chemical Kinetics*, McGraw-Hill Book Company, New York, 1960.

4.4 The concept of the rate-determining active center is not new. However, the name "rate-determining step" or "slow step" is sometimes used to denote the irreversible step in which the rate-determining active center is reacted. This practice leads to confusion, as shown by the fact that in the hydrogen-bromine reaction at high conversion, the rate-determining step is the one that involves hydrogen atoms, although bromine atoms are clearly the rate-determining active centers. The concept of rate-determining active centers has been used implicitly by N. N. Semenov, *Problems in Chemical Kinetics and Reactivity*, Vols. I and II, translated by M. Boudart, Princeton University Press, Princeton, N.J., 1958.

Coupled Sequences in Reaction Networks

5

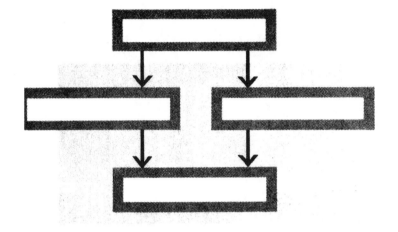

In nature and industry it happens most frequently that a reaction is not carried out with a pure feed but with a mixture of two or more related compounds that react in parallel or in series with a common reactant or on the same catalyst. More than one single reaction must then be considered, and we have to deal with a reaction network. In order to illustrate the physical ideas, a network of only two reactions will be treated, but generalization is only a matter of patience. Let us examine in turn the case of catalytic reactions and that of chain reactions. There are two problems. The first one is a problem where time does not enter explicitly — it is a problem of *selectivity*. The second problem is the kinetic problem itself: What is the total rate when both reactions go side by side?

5.1 Catalytic Reactions in Parallel

For simplicity we will consider only irreversible reactions. Two reactants A_1 and A_2 may react on a catalyst with a common reactant B to give some products which need not be specified:

$$A_1 + B \rightarrow \text{Products} \quad (1)$$

$$A_2 + B \rightarrow \text{Products} \quad (2)$$

The rate of the first reaction will be given by an expression of the form:

$$r_1 = \frac{k_1 K_1 (A_1) K_B (B)}{f_1(c)} \tag{5.1.1}$$

For the second reaction:

$$r_2 = \frac{k_2 K_2 (A_2) K_B (B)}{f_2(c)} \tag{5.1.2}$$

The constants K_1, K_2 and K_B are adsorption equilibrium constants for A_1, A_2 and B respectively while k_1 and k_2 are rate constants for the rate-determining step at the surface, involving A_1 and A_2.

The two functions of concentration $f_1(c)$ and $f_2(c)$ will in general be different if both reactions are run separately. But if they are run together, these functions will be the same: $f_1(c) = f_2(c) = f(c)$. There is no need to specify these functions any further for the problem of selectivity. The only essential assumption behind (5.1.1) and (5.1.2) is that the catalytic reaction has a rate-determining step which consists in a surface reaction of adsorbed A_1 (or A_2) which is first-order with respect to the active center involving A_1 (or A_2). The mode of participation of B in this rate-determining step is again immaterial for the problem of selectivity. In fact, B might be missing altogether.

Indeed, if relative rates are what matters, any reference to the role of B or the nature of the function $f(c)$ disappears through dividing (5.1.1) by (5.1.2) side by side:

$$\frac{r_1}{r_2} = \frac{k_1 K_1}{k_2 K_2} \frac{(A_1)}{(A_2)} \tag{5.1.3}$$

Since the stoichiometric coefficients of A_1 and A_2 are unity:

$$r_1 = -\frac{d(A_1)}{dt} \qquad r_2 = -\frac{d(A_2)}{dt}$$

Therefore, (5.1.3) becomes

$$\frac{d(A_1)}{d(A_2)} = \frac{k_1 K_1}{k_2 K_2} \frac{(A_1)}{(A_2)} \tag{5.1.4}$$

Integration of (5.1.4) between zero time and time t with corresponding concentrations indicated by subscripts 0 and t, gives:

$$S = \frac{k_1K_1}{k_2K_2} = \frac{\log (A_1)_0 - \log (A_1)_t}{\log (A_2)_0 - \log (A_2)_t} \qquad (5.1.5)$$

where the ratio of the products of rate constants and adsorption equilibrium constants for the competing reactants has been designated by S, a constant characteristic of the selectivity of the process. This constant should not depend on the extent of the competing reactions and should be equal to the function of concentrations shown on the right-hand side of (5.1.5). The *selectivity* S can be called the *reactivity ratio* of A_1 and A_2 on the catalyst being used. It is interesting to note that *relative reactivity* is not defined by a rate cosntant alone but by the product of a rate constant and an adsorption equilibrium constant, e.g., k_1K_1 for reactant A_1. This is yet another illustration of the relevance of thermodynamics to the kinetics of chemical change.

Problem 5.1.1

On a nickel catalyst at 170°C, in the liquid phase and under high pressures of hydrogen, the following values of selectivity have been obtained (the first compound is A_1, the second A_2):

	S
tetraline to paraxylene	2.80
benzene to tetraline	6.60
benzene to orthoxylene	14.5

What is the selectivity for competitive hydrogenation of mixtures of ortho and paraxylene?

5.2 Catalytic Reactions in Series

In the first example, we saw how the individual catalytic sequences $A_1 + B \rightarrow$ Products and $A_2 + B \rightarrow$ Products were coupled as a result of the competition of A_1 and A_2 for the same active centers. In general, coupling may affect profoundly the selectivity of the process.

This is also true if A_1 and A_2 react not in parallel but in series. Consider two catalytic reactions in series:

$$A_1 \rightarrow A_2 \qquad (1)$$

$$A_2 \rightarrow A_3 \qquad (2)$$

Assume that A_1 is very strongly adsorbed on the catalyst surface but that A_2 is only weakly adsorbed while A_3 is not adsorbed at all. If reactions (1) and (2) were uncoupled, their rates would be given for instance by the expressions:

$$r_1 = \frac{k_1 K_1 (A_1)}{1 + K_1(A_1) + K_2(A_2)} \cong k_1 \tag{5.2.1}$$

$$r_2 = \frac{k_2 K_2 (A_2)}{1 + K_2(A_2)} \cong k_2 K_2 (A_2)$$

since the assumptions made above are equivalent to the inequalities $K_1(A_1) \gg 1 \gg K_2(A_2)$.

The selectivity of the process would then be given by the relation giving the maximum value of (A_2):

$$r_1 - r_2 = \frac{d(A_2)}{dt} = k_1 - k_2 K_2 (A_2) = 0$$

Thus:

$$(A_2)_{max} = \frac{k_1}{k_2 K_2} \tag{5.2.2}$$

Of course this treatment is incorrect because both sequences are coupled and instead of (5.2.2) the true expression for r_2 is

$$r_2 = \frac{k_2 K_2 (A_2)}{1 + K_1(A_1) + K_2(A_2)} \cong k_2 \frac{K_2}{K_1} \frac{(A_2)}{(A_1)} \tag{5.2.3}$$

Consequently, at the maximum concentration of (A_2):

$$\frac{d(A_2)}{dt} = k_1 - k_2 \frac{K_2}{K_1} \frac{(A_2)}{(A_1)} = 0$$

Thus:

$$\frac{(A_2)_{max}}{(A_1)} = \frac{k_1 K_1}{k_2 K_2} \tag{5.2.4}$$

Comparing (5.2.2) and (5.2.4), we see that the maximum concentration of A_2, a measure of the selectivity of the process, is $K_1(A_1)$ times larger in the case where coupling occurs than in the case where it does not. Since by assumption $K_1(A_1) \gg 1$, the gain in selectivity due to coupling is most important and again it is essentially thermodynamic in nature.

In particular, A_1 may be so much more strongly adsorbed than A_2 that the surface is totally unavailable to A_2 until A_1 has completely disappeared from the reacting system. Then the selectivity of the process would have its highest possible value, i.e., the process would be quantitative; A_1 would be transformed with 100% selectivity to A_2 in a time $t = [k_1/(A_1)_0]$ at a constant rate $r_1 = k_1$. Such cases are known.

As an illustration, when a gaseous mixture of 2-butyne and deuterium is passed over a palladium catalyst at 14°C, the composition of the hydrocarbon fraction in the product is as shown in Table 5.2.1. In this example, A_1 is 2-butyne, A_2 is *cis*-2-butene-2,3-d_2 and A_3 is butane. The selectivity to butene is 99.9%. The stereoselectivity to *cis*-2-butene is 99%. The quantitative yield of butene is due to the occupancy of the surface by butyne. As long as there remains in the system enough unconverted butyne, butene has no access to the surface for its further hydrogenation to butane. Yet, all the butene will be hydrogenated readily to butane in the absence of butyne.

Thus we see how the first and foremost problem of reactions in series — that of selectivity — can be controlled by the coupling between the reaction. For catalytic sequences in parallel, it is worth while to consider also briefly the second problem, i.e., the kinetic problem itself. It consists in finding out the effect of the coupling on the overall rate r_0 of the process $r_0 = \sum_i r_i$. If there is coupling, r_0 *will not normally be the sum of the rates of the i stoichiometrically simple*

Table 5.2.1

MOLE FRACTION OF HYDROCARBONS IN PRODUCT OF THE REACTION BETWEEN 2-BUTYNE AND DEUTERIUM AT 14°C*

2-Butyne:	$CH_3—C≡C—CH_3$	0.220
cis-2-Butene-2,3-d_2:	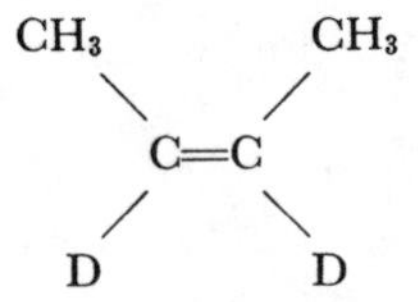	0.772
trans-2-Butene-2,3-d_2:	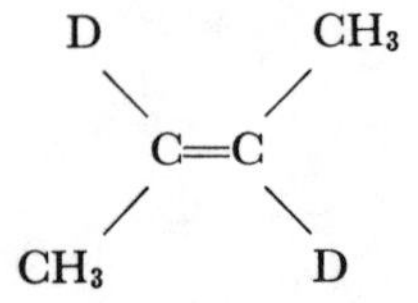	0.007
1-Butene:	$CH_2=CH—CH_2—CH_3$	0.000
Butane:	$CH_3—CH_2—CH_2—CH_3$	0.001

*From Edwin F. Meyer and Robert L. Burwell, Jr., *J. Am. Chem. Soc.*, **85**, 2877 (1963)

reactions taken individually. In particular, although each of the individual rates r_i will decrease as its own extent of reaction X_i increases, the overall rate may occasionally increase with time.

In order to show this behavior, let us return to the example treated in the preceding section; two reactants A_1 and A_2 reacting in parallel on a catalyst with a common reactant B which will be assumed to be missing from the rate expressions. These are for the case of saturation of the surface and negligible adsorption of products:

$$r_1 = k_1 \tag{5.2.5}$$

$$r_2 = k_2 \tag{5.2.6}$$

These are valid only if the two reactions proceed separately. If they take place in parallel, (5.2.5) and (5.2.6) become:

$$r_1 = \frac{k_1 K_1(A_1)}{K_1(A_1) + K_2(A_2)} \tag{5.2.7}$$

$$r_2 = \frac{k_2 K_2(A_2)}{K_1(A_1) + K_2(A_2)} \tag{5.2.8}$$

Consequently, the overall rate $r_0 = r_1 + r_2$ is quite different from the value $k_1 + k_2$ that would prevail in the absence of coupling. In particular, suppose that A_1 alone reacts faster than A_2 alone, i.e., $k_1 > k_2$. On the other hand, A_2 is more strongly adsorbed than A_1, i.e. $K_2 > K_1$. Then, in a mixture of A_1 and A_2, the slow component will react selectively at first since it competes more favorably for the active centers. As the reacting mixture becomes poorer in the slow component, the surface becomes more available to the fast component so that the overall rate increases with time.

Problem 5.2.1

Referring back to the situation described in Problem 5.1.1, consider the cohydrogenation of tetraline A_1 and *para*xylene in a molar ratio of 2 to 1. The system can be described by the rate expressions just discussed. Rate constants in g-mole/g catalyst/minute are:

$$k_1 = 6.7 \times 10^{-3} \qquad \text{and} \qquad k_2 = 12.9 \times 10^{-3}$$

Calculate overall rates during the entire course of reactions, the initial rate being taken as unity. Neglect minor volume changes during reaction. Plot, versus time, the overall rate as well as the composition of the system.

5.3 Chain Reactions in Parallel

Coupling of catalytic reactions is caused by competition for active centers. Since chain reactions are essentially catalytic in nature, coupling should also affect their rates and selectivity. But, as mentioned before, chain reactions have some characteristics that warrant a separate treatment. This treatment will now be undertaken for reactions both in parallel and in series. As examples of coupling between chain reactions running in parallel, we shall examine copolymerization and co-oxidation. The partial oxidation of methane to formaldehyde and carbon monoxide will be taken up in the next section as a typical case of coupling between chain reactions in series.

Let us first consider a special type of chain transfer (see Section 3.6) as it occurs in the simultaneous addition polymerization of two monomers M_1 and M_2. This is the case of *copolymerization*. Only the problem of selectivity that determines the composition of the copolymer will be treated here. The situation is best represented by the following scheme:

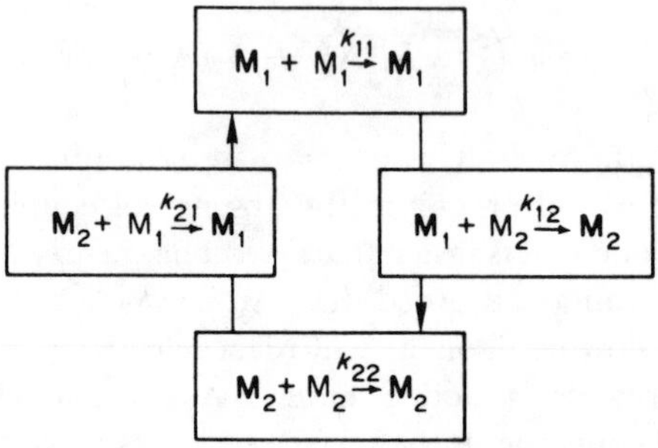

(5.3.1)

The scheme shows above and below the two propagation steps that would take place in polymerization of M_1 and M_2 separately. On the sides, left and right, are the chain transfer steps that establish coupling between the chains in copolymerization. Rate constants k_{ij} are indicated, referring to addition of monomer M_j to a growing chain $\mathbf{M}_i$, i.e., one which has last added a monomer unit M_i.

Now at the steady state, the rate at which $\mathbf{M}_2$ becomes $\mathbf{M}_1$ (on the left) must be equal to the rate at which $\mathbf{M}_1$ becomes $\mathbf{M}_2$ (on the right) since otherwise the steady-state conditions $d(\mathbf{M}_1)/dt = 0$ and $d(\mathbf{M}_2)/dt = 0$ would be violated. In other words, the chain transfer steps must proceed at the same rate:

$$k_{21}(\mathbf{M}_2)(\mathrm{M}_1) = k_{12}(\mathbf{M}_1)(\mathrm{M}_2) \tag{5.3.2}$$

The rates of disappearance of monomer are:

$$-\frac{d(M_1)}{dt} = k_{11}(\mathbf{M}_1)(M_1) + k_{21}(\mathbf{M}_2)(M_1) \tag{5.3.3}$$

$$-\frac{d(M_2)}{dt} = k_{12}(\mathbf{M}_1)(M_2) + k_{22}(\mathbf{M}_2)(M_2) \tag{5.3.4}$$

Dividing (5.3.3) and (5.3.4) side by side and using the steady-state relation (5.3.2) we obtain the basic equation of copolymerization:

$$\frac{d(\mathbf{M}_1)}{d(\mathbf{M}_2)} = \frac{(M_1)}{(M_2)}\frac{\rho_1(\mathbf{M}_1) + (M_2)}{\rho_2(\mathbf{M}_2) + (M_1)} \tag{5.3.5}$$

where the *monomer reactivity ratios*

$$\rho_1 = \frac{k_{11}}{k_{12}} \quad \text{and} \quad \rho_2 = \frac{k_{22}}{k_{21}} \tag{5.3.6}$$

have been introduced. These play in copolymerization the role of relative volatility in distillation and there is indeed a great analogy between the two processes. This can be seen by introducing mole fractions x_1 and x_2 of monomers M_1 and M_2 and y_1, the mole fraction of M_1 in the polymer. The basic equation becomes:

$$y_1 = \frac{\rho_1 x_1^2 + x_1 x_2}{\rho_1 x_1^2 + 2x_1 x_2 + \rho_2 x_2^2} \tag{5.3.7}$$

If $\rho_1\rho_2 = 1$, the copolymerization is said to be ideal: The probabilities that M_1 follows M_2 and M_2 follows M_1 are the same and y_1 reduces to:

$$y_1 = \frac{\rho_1 x_1}{\rho_1 x_1 + x_2} \tag{5.3.8}$$

If $\rho_1\rho_2 \neq 1$, the copolymerization is nonideal; the smaller this product, the more regular is the alternation of monomers in the polymer chain. Alternation is perfect ($y_1 = \frac{1}{2}$) at all concentrations of monomer when $\rho_1 = \rho_2 = 0$, i.e., when the monomers do not polymerize separately, as for instance maleic anhydride and stilbene.

In general, of course, the composition of the copolymer will change as reaction proceeds. As in distillation, however, there are azeotropic systems for which the composition of the copolymer will be the same as that of the reacting system. The condition for azeotropy, $y_1 = x_1$, is satisfied when

$$x_1 = \frac{1 - \rho_2}{2 - \rho_1 - \rho_2} \tag{5.3.9}$$

if both ρ_1 and ρ_2 are smaller or larger than unity.

The second problem is here again the effect of the coupling of the overall rate of two chain reactions in parallel. As an example, consider the liquid phase co-oxidation of two hydrocarbons R_1H and R_2H, at sufficiently high pressures of oxygen so that the peroxy radicals R_1OO and R_2OO are the rate-determining active centers (see page 101). Then, the sequence of propagation is completely defined by the following scheme:

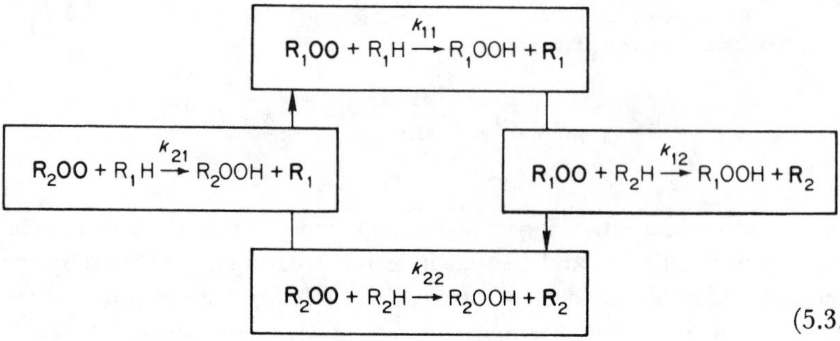

(5.3.10)

There is no need to write the steps

$$\mathbf{R_1} + O_2 \rightarrow \mathbf{R_1OO} \quad \text{and} \quad \mathbf{R_2} + O_2 \rightarrow \mathbf{R_2OO}$$

since they are not kinetically significant. But of course the three possible termination steps must be included:

$$\left.\begin{aligned} \mathbf{R_1OO} + \mathbf{R_1OO} &\xrightarrow{k'_{11}} \\ \mathbf{R_1OO} + \mathbf{R_2OO} &\xrightarrow{k'_{12}} \\ \mathbf{R_2OO} + \mathbf{R_2OO} &\xrightarrow{k'_{22}} \end{aligned}\right\} \text{inert molecules} \tag{5.3.11}$$

Also the rate of initiation r_i must be specified, although the nature of the initiation step remains immaterial. The condition of steady state will be expressed, as in the case of copolymerization, by the equality of the rates of the two chain transfer steps, as shown on the scheme on the left and right:

$$k_{21}(\mathbf{R_2OO})(R_1H) = k_{12}(\mathbf{R_1OO})(R_2H) \tag{5.3.12}$$

This condition, together with the usual equality between rates of termination and initiation, gives readily the expression for the overall rate of oxidation, equal to the sum of the rates of the four steps shown in the scheme. The result is:

$$r_0 = r_i^{1/2} \frac{\rho_1(R_1H)^2 + 2(R_1H)(R_2H) + \rho_2(R_2H)^2}{[\alpha_1(R_1H)^2 + 2\beta(R_1H)(R_2H) + \alpha_2(R_2H)^2]^{1/2}} \tag{5.3.13}$$

where

$$\rho_1 = \frac{k_{11}}{k_{12}} \qquad \rho_2 = \frac{k_{22}}{k_{21}} \tag{5.3.6}$$

as in copolymerization, and

$$\tfrac{1}{2}\alpha_1 = \frac{k'_{11}}{k_{12}^2} \qquad \beta = \frac{k'_{12}}{k_{12}k_{21}} \qquad \tfrac{1}{2}\alpha_2 = \frac{k'_{22}}{k_{21}^2} \tag{5.3.14}$$

It is interesting to compare the value of the overall rate (5.3.13) to that which would obtain if coupling were absent:

$$r_0 = r_i^{1/2}\left[\frac{\rho_1}{\alpha_1^{1/2}}(R_1H) + \frac{\rho_2}{\alpha_2^{1/2}}(R_2H)\right] \tag{5.3.15}$$

Not only is the form of the rate expression changed appreciably by the chain transfer reactions, but also the value of the overall rate may suffer changes that are not immediately obvious. These changes are due to the great influence of rates of termination on the rates of chain reactions, and also to the effect of coupling.

Consider for instance the effect on the overall rate of the addition to pure R_1H of a small quantity of R_2H. Will the rate increase or decrease? Will R_2H act as an accelerator or inhibitor? With $[(R_2H)/(R_1H)] = z$, let us rewrite the equation for the overall rate:

$$r = r_i^{1/2}(A_1)\frac{\rho_1 + 2z + \rho_2 z^2}{[\alpha_1 + 2\beta z + \alpha_2 z^2]^{1/2}}$$

The partial derivative of r with respect to z at $z = 0$ is proportional to:

$$2\alpha_1 - \rho_1\beta$$

It will be positive or negative according to whether

$$\frac{2}{\rho_1} - \frac{\beta}{\alpha_1}$$

is positive or negative. Thus, R_2H will be an accelerator if

$$4\frac{k_{21}}{k_{11}} > \frac{k'_{12}}{k'_{11}} \tag{5.3.16}$$

i.e., when four times the ratio of cross-propagation to self-propagation is larger than the ratio of cross-termination to self-termination.

Conversely, R_2H will be an inhibitor when

$$\frac{k'_{12}}{k'_{11}} > 4\frac{k_{21}}{k_{11}} \tag{5.3.17}$$

It is seen that the kinetic behavior of R_2H alone, as determined by its own rate constants for self-propagation and self-termination, k_{22} and k'_{22}, do not appear in these inequalities. Thus it may happen that a substance R_2H which, by itself, reacts faster than R_1H will nevertheless depress the overall rate of reaction when added in small quantity to pure R_1H.

Problem 5.3.1

Verify the rate expression (5.3.13).

Problem 5.3.2

For the oxidation of the system cumene (R_1H)–tetraline (R_2H), at 90°C and sufficiently high pressures of oxygen, the significant rate constants are (in liter/mole-sec):

$$k_{11} = 1.3 \qquad k'_{11} = 4.2 \times 10^4$$

$$k_{12} = 30 \qquad k'_{12} = 6 \times 10^6$$

$$k_{21} = 3.2 \qquad k'_{22} = 2.6 \times 10^7$$

$$k_{22} = 51$$

Find whether tetraline will act as an inhibitor or accelerator when added in small amounts to cumene. Conversely, find the effect on the rate of a small addition of cumene to tetraline. Other things being equal, which pure compound oxidizes faster, cumene or tetraline?

5.4 Chain Reactions in Series

The last question that will be examined on the subject of chain transfer is the effect of coupling between chain reactions in series on the selectivity of the system. The example chosen is the gas-phase oxidation of methane at controlled temperatures. The two consecutive reactions that are important in the early stages of the reaction are:

$$CH_4 + O_2 \rightarrow CH_2O + H_2O \qquad (1)$$

$$CH_2O + O_2 \rightarrow CO + H_2O_2 \qquad (2)$$

These would be followed at later stages by two more:

$$CO + \tfrac{1}{2}\,O_2 \rightarrow CO_2$$

$$H_2O_2 \rightarrow H_2O + \tfrac{1}{2}\,O_2$$

which will not be considered here.

Both reactions (1) and (2) are chain reactions and they are coupled according to the following scheme:

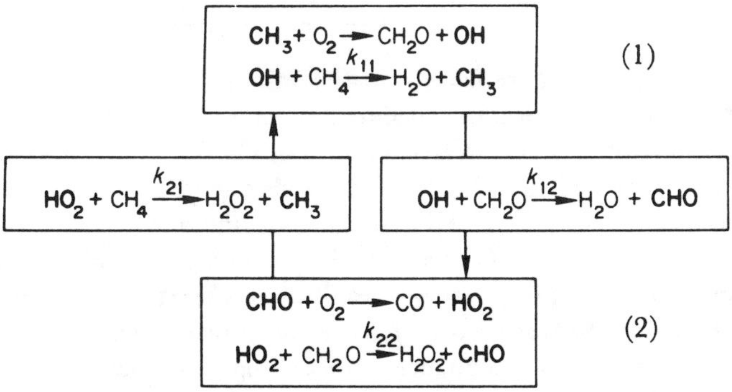

The analogy with copolymerization appears clearly with $M_1 \equiv CH_4$, $M_2 \equiv CH_2O$, $\mathbf{M}_1 \equiv OH$, $\mathbf{M}_2 \equiv HO_2$. At the steady state the chain transfer steps which couple both chains (1) and (2) above and below, must proceed at the same rate. Thus:

$$a_{12}(OH) = a_{21}(HO_2) \qquad (5.4.1)$$

Both sequences (1) and (2) will normally proceed at different rates. For two reactions in series, the maximum amount of intermediate product (i.e., the product of the first reaction) is reached when the rates of both successive reactions become equal. This condition determines the optimum selectivity of the process. Therefore, the maximum concentration of formaldehyde will be reached when:

$$a_{11}(\mathrm{OH}) = a_{22}(\mathrm{HO_2}) \tag{5.4.2}$$

Dividing (5.4.2) by (5.4.1) side by side

$$\frac{a_{11}}{a_{12}} = \frac{a_{22}}{a_{21}} \tag{5.4.3}$$

The relation can be rewritten in the form

$$\frac{k_{11}}{k_{12}}\frac{(\mathrm{CH_4})}{(\mathrm{CH_2O})} = \frac{k_{22}}{k_{21}}\frac{(\mathrm{CH_2O})}{(\mathrm{CH_4})} \tag{5.4.4}$$

The maximum concentration of formaldehyde is given by:

$$\boxed{(\mathrm{CH_2O})_{\max} = \left(\frac{k_{11}k_{21}}{k_{22}k_{12}}\right)^{1/2}(\mathrm{CH_4})_0} \tag{5.4.5}$$

where $(\mathrm{CH_4})$ has been replaced by the initial concentration of methane $(\mathrm{CH_4})_0$ because the maximum concentration of intermediate is reached in the early stages of the reaction so that $(\mathrm{CH_4}) \sim (\mathrm{CH_4})_0$.

It must be stressed that this result depends essentially on the coupling between sequences. A study of reaction (1) alone and of reaction (2) alone with the condition that rates of (1) and (2) thus measured should be equal at the maximum, would lead to erroneous results. In all systems *where more than one stoichiometrically simple reaction takes place, coupling between sequences must be taken into account*, whether the sequences are arranged in parallel or in series, whether they are catalytic or chain reactions. Selectivity, product composition and overall rates may be affected substantially by the competition for active centers and the chain transfer steps. In other words, if two rate functions, for two reactions taking place separately, are of the form:

$$r_1 = r_1(p,T,\mathrm{X}_1) \qquad r_2 = r_2(p,T,\mathrm{X}_2)$$

they will be of the form:

$$r_1' = r_1'(p,T,\mathrm{X}_1,\mathrm{X}_2) \qquad r_2' = r_2'(p,T,\mathrm{X}_1,\mathrm{X}_2)$$

when they proceed together in a common reacting system. It is this circumstance that makes the kinetic analysis of reaction networks particularly difficult.

BIBLIOGRAPHY

5.1 This topic has been developed by Jungers, Balacéanu and co-workers from whose
5.2 work Problems 5.1.1 and 5.2.1 have been drawn. See: J.C. Jungers and J.C. Balacéanu, "Investigation of Rates and Mechanisms of Reaction," Part I, Chapter XIII, *Technique of Organic Chemistry*, (2nd ed.) Vol. VIII, A. Weissberger, ed., Interscience Publishers, New York, (1961).

5.3 For an introduction to the kinetics of copolymerization, see J.C. Bevington, *Radical Polymerization*, Academic Press Inc., New York, 1961. Co-oxidation has been studied by Russell, *J. Am. Chem. Soc.*, **77**, 4583 (1955), and more recently by Alagy, Clément and Balacéanu, *Bull. Soc. Chim. France*, 1325 (1959), 1495 (1960), 1303 (1961). Problem 5.3.3. is taken from the work of the latter authors. The inequalities (5.3.19) and (5.3.20) have been derived by Tsepalov, *Zhur. Fiz. Khim.*, **35**, 1086, 1443, 1691 (1962).

5.4 The partial oxidation of methane has been studied many times. Recent work has been summarized by N.N. Semenov, *Some Problems in Chemical Kinetics and Reactivity*, Vol. II, Chapter XII (M. Boudart, trans.), Princeton University Press, Princeton, N.J., 1959.

Autocatalysis and Inhibition

6

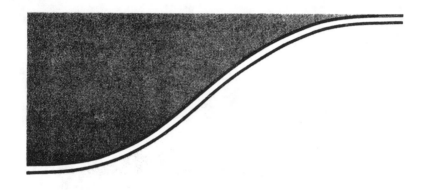

6.1 *Acceleration of the Rate*

In general, as formulated in Rule I of Chapter 1, the rate of reaction decreases as the extent of reaction increases. In the case of more than one single reaction, exceptions to this statement can occur as a result of competition for active centers (see Section 5.2). But even when the reaction is stoichiometrically simple, such an acceleration of the rate is possible if the concentration of active centers builds up as reaction proceeds. Then the steady-state approximation apparently fails: There does not seem to exist a short relaxation time beyond which the concentration of active centers ceases to depend explicitly on time (see p. 67).

Situations of this type are grouped under the term *autocatalysis*, which means that the system produces its catalyst as it goes along. As in the growth of populations, the acceleration can become so intense that the word *explosion* properly describes the catastrophic event. The explosion may fizzle if reactants are exhausted before the situation has gone out of hand — the rate will then increase at first, go through a maximum which does not have too large a value, then decrease again. The explosion is said to be *degenerate*.

Autocatalysis can be due to different causes. Its kinetic features can be recognized in a simple case. The hydrolysis of an ester RCOOR′ in a dilute water solution takes place at a very slow rate according to:

$$\mathrm{RCOOR' + H_2O \rightarrow RCOOH + R'OH} \qquad (1)$$

This is called the "residual" reaction in the absence of any catalyst. The rate is first-order with respect to the ester and zero-order with respect to the water because the latter is present in large excess so that its concentration stays constant for all practical purposes. Therefore, at constant volume:

$$r_1 = -\frac{d(\mathrm{RCOOR'})}{dt} = k_1(\mathrm{RCOOR'}) \qquad (6.1.1)$$

If a is the initial concentration of ester and f the fraction converted, (6.1.1) becomes:

$$-\frac{d[a(1-f)]}{dt} = k_1 a(1-f)$$

or:

$$\frac{df}{dt} = k_1(1-f) \qquad (6.1.2)$$

The acid RCOOH produced in the hydrolysis is however a catalyst for the reaction itself and, in parallel, a second reaction with a much larger rate constant takes place at a rate r_2, proportional to the concentration of catalyst. Thus:

$$r_2 = -\frac{d(\mathrm{RCOOR'})}{dt} = k_2(\mathrm{RCOOR'})(\mathrm{RCOOH}) \qquad (6.1.3)$$

The overall rate is then:

$$r_0 = r_1 + r_2 = k_1(\mathrm{RCOOR'}) + k_2(\mathrm{RCOOR'})(\mathrm{RCOOH})$$

Thus:

$$\frac{df}{dt} = (k_1 + k_2 af)(1-f) \qquad (6.1.4)$$

The constant $k = k_2a$ is a bimolecular rate constant times initial concentration. Its inverse is a measure of the reaction time for the catalytic reaction. Similarly, k_1 is a pseudo first-order rate constant: its inverse is a measure of the reaction time for the residual reaction. Consequently, the ratio $\rho = k_1/k$ is a fraction and its value is very small compared to unity (say $\rho \sim 0.01$). Then (6.1.4) becomes:

$$\frac{df}{dt} = k(f + \rho)(1 - f) \tag{6.1.5}$$

This expression, first derived by Ostwald in 1883, exhibits the main features of autocatalytic behavior. Integrating (6.1.5) with $f = 0$ at $t = 0$, we get:

$$f = \rho \frac{\exp[(1 + \rho)kt] - 1}{1 + \rho \exp[(1 + \rho)kt]} \tag{6.1.6}$$

Let $\tau = (1 + \rho)kt$ be a dimensionless time. Then:

$$\boxed{f = \rho \frac{1 - \exp(-\tau)}{\rho + \exp(-\tau)}} \tag{6.1.7}$$

The curve of conversion versus time first rises showing autoacceleration, goes through an inflection point, then tapers off so that $f \to 1$ as $\tau \to \infty$. The curve shown in Fig. 6.1.1 looks like an integral sign $\int$ and is said to have a characteristic S-shape. The rate reaches its maximum value at a time $\tau = -\ln \rho$ corresponding to the inflection point in the f-curve. At this time both conversion f and the dimensionless maximum rate $(2/k)\, df/dt$ are equal to $\frac{1}{2}$.

Problem 6.1.1

More generally, the rate equation for an autocatalytic reaction has the empirical form:

$$\frac{df}{dt} = k(f + \rho)^{\alpha}(1 - f)^{\beta}$$

At what degree of conversion is the maximum rate reached if $\alpha = 1$ and $\beta = 2$?

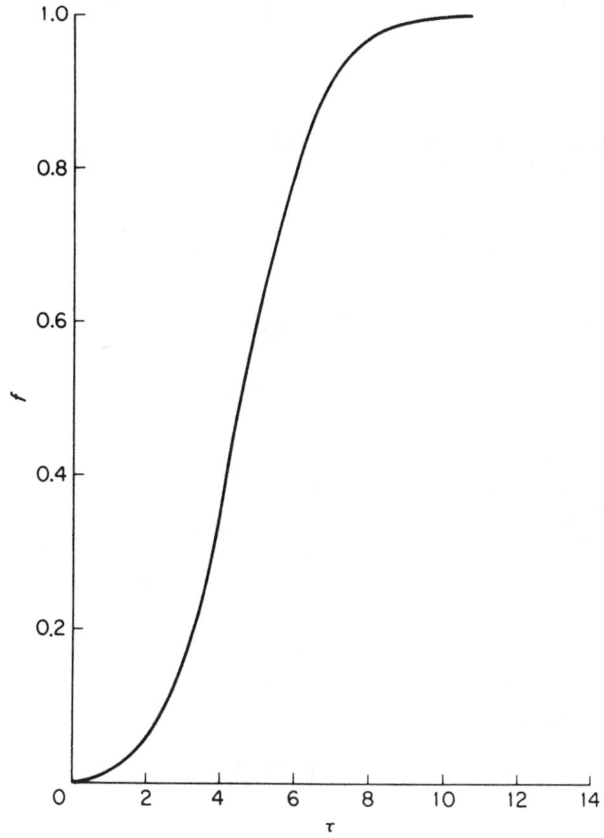

Fig. 6.1.1 The autocatalytic curve, a plot of degree of conversion f versus dimension less time τ for rate described by Eq. (6.1.5).

6.2 *Branched Chains*

The chain reactions treated in Section 3.4 consist of straight chains. There is an initiation step taking place at a rate r_i. The propagation steps do not produce nor do they destroy active centers. This is so because in each step of the closed sequence, the active centers change only in kind, not in number. Therefore, there exists a steady state defined by the condition that the rate of initiation is equal to the rate of termination.

Before considering the case of branched chains for which no steady state exists, it is instructive to examine the approach to steady state in the case of straight chains. This will be done for linear termination, so called because the rate of termination is first-order with respect to the rate-determining active center **X**:

$$r_t = a_t(\mathbf{X}) \tag{6.2.1}$$

At time zero, the concentration of active centers is equal to zero. It grows according to the law:

$$\frac{d(\mathbf{X})}{dt} = r_i - a_t(\mathbf{X}) \tag{6.2.2}$$

It reaches its steady-state value $(\mathbf{X})^*$ when $d(\mathbf{X})/dt = 0$. Thus:

$$(\mathbf{X})^* = \frac{r_i}{a_t} \tag{6.2.3}$$

Integration of (6.2.2) shows that $(\mathbf{X})$ approaches its steady-state value $(\mathbf{X})^*$ exponentially:

$$(\mathbf{X}) = (\mathbf{X})^*[1 - \exp(-a_t t)] \tag{6.2.4}$$

This approach is usually quite rapid. Indeed, we can use again (see p. 66) the parameter ϵ expressing the departure from steady-state concentration:

$$(\mathbf{X}) = (\mathbf{X})^* (1 + \epsilon) \tag{3.2.11}$$

Then:

$$\epsilon = -\exp(-a_t t) \tag{6.2.5}$$

Comparing (6.2.5) with (3.2.20), we see that the *relaxation time* t_r which is a good measure of the time required to reach steady state is thus simply equal to $1/a_t$:

$$t_r = \frac{1}{a_t} \tag{6.2.6}$$

For a typical termination rate constant, $k_t = 10^{-13}$ cm^3/sec; if the stable molecular species M involved in the termination step (M + **X** $\rightarrow$ stable products) is at a partial pressure as low as 1 torr, (M) $\cong 10^{16}$ cm^{-3} in order of magnitude. Consequently, $a_t = 10^{-13} \cdot 10^{16} = 10^3$ sec^{-1} and the relaxation time is of the order of one millisecond. This figure is given here only for the sake of illustration. Of course other situations would give other figures and also, it may happen that the time of relaxation becomes comparable to the time of reaction.

Another characteristic time which is frequently used in connection with chain reactions is the *induction period* t_i. It must not be confused with the relaxation time. During the induction period, the rate of reaction proceeds so slowly that no change in the reacting system is observable with a given analytical tool. The definition of t_i is somewhat arbitrary since it depends on the power of resolution of the analytical device. For instance, if it is possible to detect active centers by spectroscopy, the threshold of sensitivity of the apparatus would correspond to a concentration of active centers $(\mathbf{X})_m$ where m stands for measurable. Then the induction period would be defined by means of the relation (6.2.4) which, together with (6.2.6), gives:

$$(\mathbf{X})_m = (\mathbf{X})^*\left[1 - \exp - \left(\frac{t_i}{t_r}\right)\right] \tag{6.2.7}$$

The value of t_i is then:

$$t_i = t_r \ln\left[1 + \frac{(\mathbf{X})_m}{(\mathbf{X})^*}\right] \tag{6.2.8}$$

Let us now consider a chain reaction with the usual steps for initiation, propagation and termination, but in addition having a step of a new type in which more than one active center (two, for example) are produced from only one active center. A typical step of this kind is:

$$\mathbf{H} + \mathrm{O_2} \quad \rightarrow \quad \mathbf{OH} + \mathbf{O}$$

This is called a *branching* step and the chain reaction is now called a *branched-chain reaction:* Each of the two active centers will start its own chain so that two chains are created where only one existed before. Whereas the propagation steps need not be taken into account in the differential equation describing the increase of $(\mathbf{X})$ with time, the effect of the branching step must be included. If the rate of the branching step is $r_b = a_b(\mathbf{X})$, the concentration of active centers changes with time following the relation:

$$d(\mathbf{X})/dt = r_i + a_b(\mathbf{X}) - a_t(\mathbf{X}) \tag{6.2.9}$$

Let φ denote the difference $a_b - a_t$. Then:

$$d(\mathbf{X})/dt = r_i + \varphi(\mathbf{X}) \tag{6.2.10}$$

If the quantity φ, called the *net branching factor*, is negative, the situation is identical to that just described for the case of straight chains by Eq. (6.2.2): There exists a steady state and the concentration of active centers approaches exponentially its steady-state value $r_i/|\varphi|$.

If on the other hand the net branching factor φ is positive, integration of (6.2.10) now yields:

$$(\mathbf{X}) = \frac{r_i}{\varphi}(e^{\varphi t} - 1) \tag{6.2.11}$$

The concentration of active centers increases exponentially with time. When $\exp(\varphi t) \gg 1$ (6.2.11) becomes:

$$\boxed{(\mathbf{X}) = \left(\frac{r_i}{\varphi}\right)\exp(\varphi t)} \tag{6.2.12}$$

No steady-state reaction is possible. The reaction rate, which is proportional to (**X**) also increases exponentially with time; the autocatalysis reaches catastrophic proportions and explosion takes place. It must be stressed that the primary cause of the explosion is not the accumulation of heat in the system, as occurs in thermal explosions (see Chapter 7). The self-acceleration of the rate can take place isothermally. Naturally, as the reaction rate becomes very high, self-heating of the system may also take place and contribute to the explosion.

In summary, when the net branching factor φ is negative, the reaction proceeds at a steady-state rate, but when it is positive, explosion occurs. *The relation $\varphi = 0$ then represents a critical condition for explosion.* If the rate of initiation r_i is very small and if the rate of branching is sufficiently high, the critical condition is exceedingly sharp, i.e., a very small change in the variables that determine φ will produce explosion. This is shown by a numerical argument due to Semenov who first pointed out the significance of branched-chain reactions.

Suppose that $r_i = 10$ molecules/cm³-sec and $a_p = a_b = 5 \times 10^2$ sec^{-1}. The latter figures mean that the average lifetime of an active center before it reacts in either the propagation step or the branching step is 2×10^{-3} sec. Let us see what happens when the ratio (a_t/a_b) is changed by only 1% on either side of the value $(a_t/a_b) = 1$ corresponding to the critical condition $\varphi = a_b(1 - a_t/a_b) = 0$. When $(a_t/a_b) = 1.01$, $\varphi = -5$ sec^{-1} but when $(a_t/a_b) = 0.99$, $\varphi = +5$ sec^{-1}.

When $\varphi = -5$, the steady-state rate is equal to

$$r = a_p(\mathbf{X})^* = a_p\frac{r_i}{|\varphi|} = 10^3 \text{ molecules/cm}^3\text{-sec} \tag{6.2.13}$$

This figure is so ridiculously low that no detectable change would occur in the system for an extremely long time.

By contrast, when $\varphi = 5$, after only 5 seconds, the exponentially increasing rate has reached a very high value:

$$r = a_p(\mathbf{X}) = a_p \frac{r_i}{\varphi} e^{25} \cong 10^{14} \text{ molecules/cm}^3\text{-sec},$$

a figure which is 10^{11} times higher than the preceding one. Thus, the critical condition $\varphi = 0$ is truly a very sharp one.

Problem 6.2.1

In a stoichiometric hydrogen-oxygen mixture at 800°K, there exists a so-called second explosion limit which is reached at a critical pressure determined by the branching step

$$\mathbf{H} + O_2 \xrightarrow{k_b} \mathbf{OH} + \mathbf{O}$$

and the termination step

$$\mathbf{H} + O_2 + M \xrightarrow{k_t} HO_2 + M$$

The species HO_2 is under these conditions unable to continue the chain, i.e., it is not an active center. The rate constants are:

$$k_b \cong 10^{-10} \exp\left(-\frac{E}{RT}\right) \text{cm}^3/\text{sec with } E \cong 15 \text{ kcal/g-mole}$$

$$k_t \cong 10^{-32} \text{ cm}^6/\text{sec}$$

Calculate the second explosion limit (in torr).

It must be noted that there is a net gain of *two* hydrogen atoms each time the branching step proceeds. Indeed it is followed up quickly by the propagating steps:

$$\mathbf{OH} + H_2 \rightarrow H_2O + \mathbf{H} \qquad (1)$$

$$\mathbf{O} + H_2 \rightarrow \mathbf{OH} + \mathbf{H} \qquad (2)$$

Adding up step (1) taken twice, step (2) and the equation for the branching step, it is seen that the net result of one branching act is:

$$\mathbf{H} + 3\,H_2 + O_2 \rightarrow \mathbf{H} + 2\,\mathbf{H} + 2\,H_2O$$

A little reflection will then help in setting up the proper critical condition for this system.

Problem 6.2.2

At the rate defined by Eq. (6.2.13), how long would it take to accumulate one micromole in a reactor of one liter?

6.3 Degenerate Branching

Very frequently, the phenomena described in the previous section are observed qualitatively but with considerably less sharpness, in the kind of autocatalysis associated with degenerate chain branching. Here, the active center involved in the chain branching step is not an active center at all but a relatively unstable intermediate product which, upon its decomposition or reaction provides active centers at a rate considerably faster than that of the original initiation. Thus the autocatalytic behavior can really be ascribed to a secondary initiation brought about by an intermediate product. This phenomenon happens frequently in the oxidation of hydrocarbons RH. At low temperatures, it is called autoxidation and it is autocatalytic because of the further decomposition into free radicals of hydroperoxides ROOH which are first produced in the oxidation (see p. 101).

At higher temperatures, aldehydes play the role of degenerate chain branching agents and the case chosen for illustration will again be the oxidation of methane (see Section 5.4). The primary initiation in a mixture of methane and oxygen is due to a very slow step:

$$CH_4 + O_2 \rightarrow \mathbf{CH_3} + \mathbf{HO_2}$$

This takes place at a rate:

$$r_i = k_i(CH_4)(O_2) \tag{6.3.1}$$

The active centers $\mathbf{CH_3}$ produced in this initiation propagate a closed sequence:

$$\mathbf{CH_3} + O_2 \rightarrow CH_2O + \mathbf{OH} \qquad (1)$$

$$\mathbf{OH} + CH_4 \rightarrow H_2O + \mathbf{CH_3} \qquad (2)$$

The active centers $\mathbf{HO_2}$ also participate because of the step:

$$\mathbf{HO_2} + CH_4 \rightarrow H_2O_2 + \mathbf{CH_3} \qquad (3)$$

which produces a $\mathbf{CH_3}$ active center.

But as soon as formaldehyde is produced in the closed sequence, it also is slowly attacked by oxygen:

$$CH_2O + O_2 \rightarrow \mathbf{CHO} + \mathbf{HO_2}$$

at a rate:

$$r_b = k_b(CH_2O)(O_2) = a_b(CH_2O) \tag{6.3.2}$$

which is still quite small but considerably higher than r_i. This constitutes a branching step because two active centers appear in the reaction between a stable molecule O_2 and a not too stable molecule CH_2O. Again, each of the two active centers propagates the closed sequence, $\mathbf{HO_2}$, because of step (3) and **CHO** because of the additional step

$$\mathbf{CHO} + O_2 \rightarrow CO + \mathbf{HO_2} \qquad (4)$$

also followed by step (3).

We want to show the autocatalytic behavior of the rate. But to describe the rate, we also need to specify the mode of termination. It will be assumed that termination occurs predominantly through the destruction of **OH** at a rate given by

$$r_t = a_t\,(\mathbf{OH}) \tag{6.3.3}$$

At the steady state, the kinetic chain length is equal to the ratio of rates of propagation and termination:

$$\frac{a_2(\mathbf{OH})}{a_t(\mathbf{OH})} = \frac{a_2}{a_t} \tag{6.3.4}$$

and the rate of propagation is equal to the rate of initiation times the kinetic chain length. Thus, from (6.3.2) and (6.3.4):

$$\frac{d(CH_2O)}{dt} = (2r_i + 2r_b)\frac{a_2}{a_t} = 2r_i\frac{a_2}{a_t} + 2a_b\frac{a_2}{a_t}(CH_2O) \tag{6.3.5}$$

where the factor two is due to the fact that both active centers produced in initiation or branching are able to propagate the chain. Equation (6.3.5) is of the same form as (6.2.10), with a positive net branching factor

$$\varphi = 2a_b\frac{a_2}{a_t} \tag{6.3.6}$$

Integration of (6.3.5) gives:

$$(CH_2O) = 2r_i \frac{a_2}{a_t\varphi} [e^{\varphi t} - 1] \tag{6.3.7}$$

which, for $e^{\varphi t} \gg 1$, reduces to:

$$(CH_2O) = 2r_i \frac{a_2}{a_t\varphi} e^{\varphi t} \tag{6.3.8}$$

It is seen that the intermediate product, in the very early stages of the reaction, accumulates exponentially with time. At later stages of the reaction, competition of CH_2O for the active centers $\mathbf{HO_2}$ and $\mathbf{OH}$ must be taken into account as was done in Section 5.4 when the problem of the maximum concentration of intermediate CH_2O was considered. This aspect of the problem will be taken up again soon but in the meantime we shall focus our attention on the early stages of reaction where formaldehyde is still at too low a concentration for competing effectively with methane for the active centers $\mathbf{HO_2}$ and $\mathbf{OH}$.

Under these conditions, the rate of reaction, i.e., the rate of oxidation of methane, also increases exponentially with time. Indeed even at very small degrees of conversion, r_i becomes negligible as compared to r_b. Thus, (6.3.5) simplifies to:

$$-\frac{d(CH_4)}{dt} = 2r_b \frac{a_2}{a_t} = 2a_b \frac{a_2}{a_t} (CH_2O) = \varphi(CH_2O) \tag{6.3.9}$$

Replacing (CH_2O) by its value given by (6.3.8), we get:

$$\boxed{-\frac{d(CH_4)}{dt} = \varphi 2r_i \frac{a_2}{a_t\varphi} e^{\varphi t} = 2r_i \frac{a_2}{a_t} e^{\varphi t}} \tag{6.3.10}$$

Consequently, the rate of disappearance of methane is equal to what it would be in the case of a straight-chain reaction, namely $2r_i(a_2/a_t)$, multiplied by an exponential function of t times the net branching factor.

Because the branching step is a slow one, the induction period will be protracted. In accordance with the definition given in the preceding section, the induction period t_i will be defined as the time required for the amount of methane reacted to be detectable by a suitably sensitive analytical device. Integration of (6.3.10) gives:

$$(CH_4)_0 - (CH_4)_t = \Delta(CH_4) = 2r_i \frac{a_2}{a_t\varphi} [e^{\varphi t} - 1] \tag{6.3.11}$$

Neglecting as usual unity in front of $e^{\varphi t}$ and using (6.3.6), we can write (6.3.11) more simply in the form:

$$\Delta(CH_4) = \frac{r_i}{a_b} e^{\varphi t} \tag{6.3.12}$$

The sensitivity of the analytical instrument corresponds to a certain value of $\Delta(CH_4)$, say $\Delta'(CH_4)$, reached at time t_i. Thus, (6.3.12) becomes

$$\Delta'(CH_4) = \frac{r_i}{a_b} e^{\varphi t_i} \tag{6.3.13}$$

Taking logarithms of both sides of (6.3.13) we get:

$$\ln \Delta'(CH_4) \frac{a_b}{r_i} = \varphi t_i \tag{6.3.14}$$

On the left-hand side, $\Delta'(CH_4)$ is an instrumental constant. The logarithm of (a_b/r_i) will depend on temperature and pressure but relatively weakly of course as compared to a_b and r_i themselves. Therefore, *approximately:*

$$\boxed{\varphi t_i = \text{constant}} \tag{6.3.15}$$

Frequently it is indeed observed that the product of the induction period and the net branching factor is approximately a constant in a given system and this is a useful kinetic relation.

Finally, it must be noted that the case at hand is one of degenerate branching, i.e., the explosion is degenerate — it does not take place. If formaldehyde kept on accumulating at an exponential rate, explosion would of course occur soon after the induction period was over. But, as is already known, formaldehyde will in turn react, its concentration reaches a maximum and with it, the rate will also reach a maximum value. The *maximum rate of reaction* is a very important kinetic item and a rate expression for it can be found easily. Experimentally, the maximum rate is relatively easy to determine and its value can be compared to that predicted by the rate expression.

In the case of the oxidation of methane, the maximum rate r_{max} will correspond to the maximum value of formaldehyde concentration:

$$(CH_2O)_{max} = k(CH_4)_0 \tag{6.3.16}$$

Equation (6.3.16) is identical to (5.4.5), with $k = (k_{11}k_{21}/k_{22}k_{12})^{1/2}$ in the notation used in connection with the derivation of this relation.

Then, substitution of (6.3.16) into (6.3.9) gives:

$$-\left[\frac{d(CH_4)}{dt}\right]_{max} = r_{max} = 2\,\frac{k_b k_2 k}{a_t}\,(CH_4)_0^2(O_2)_0 \qquad (6.3.17)$$

The oxygen concentration, like that of methane, is replaced by its initial value $(O_2)_0$ since, as is known, the maximum is reached at very small conversions. It is well worth noting that this rate expression has a remarkably simple form in spite of the complexity of the phenomenon. Indeed, it is once more of the serviceable type of the law of mass action, a rather unexpected result which emphasizes the great generality of our Rule IV of Chapter 1.

Problem 6.3.1

Derive a rate expression for the maximum rate of oxidation of methane, following the reaction pattern discussed above but with one exception: Assume that the rate-determining radical is $\mathbf{HO_2}$ and that its linear termination is the only pertinent termination step.

6.4 Reactions Involving Solids

Autocatalytic behavior is widespread among reactions where new solid phases are formed. An obvious example is the case of a gas decomposing into a solid and another gas, the solid formed acting as a catalyst that accelerates the decomposition of the gaseous reactant. The reaction

$$GeH_{4(g)} \rightarrow Ge_{(s)} + 2H_{2(g)}$$

is a typical illustration: If no germanium is originally present in a glass reactor, the rate of decomposition of germane will be autocatalytic because germanium deposited on the walls of the reactor acts as a catalyst for the decomposition of germane.

Conversely, if a solid is decomposed into another solid and a gas, as in many practical calcination processes, autocatalysis is frequently observed. The first observation of this kinetic phenomenon goes back to G. N. Lewis, who in 1905 studied the thermodynamic decomposition pressure of solid silver oxide. Lewis noted that it took a long time for the equilibrium pressure of oxygen to establish itself. He further noted that, once oxygen started to be evolved, the process accelerated itself. He then conducted a classical kinetic study in which he showed that his data could be well represented at all temperatures by means of Ostwald's autocatalytic rate expression (6.1.5) simplified to the form:

$$\frac{df}{dt} = kf(1 - f) \tag{6.4.1}$$

where f is the fraction of solid material having reacted at time t. This rate equation is somewhat confining since, as we have seen, it predicts that the maximum rate is reached at half-conversion, a property which is by no means generally observed in the decomposition of solids. Among the many rate equations that have been proposed to account for the S-shaped decomposition curves, there is one that appears to be particularly simple and serviceable. In its integrated form, it reads:

$$f = 1 - \exp(-kt^n) \tag{6.4.2}$$

where k and n are adjustable parameters that can easily be determined from the slope and intercept of a straight line obtained in a plot of $\log [-\log (1 - f)]$ vs $\log t$. It is frequently found that n is equal to 2 or 3, although other values are also encountered.

Many efforts have been made to derive (6.4.2) and other equations on theoretical grounds. However, no single theory will account for the many variations that are believed to occur, although (6.4.2), for instance, suffices to describe semiempirically the general autocatalytic behavior. Quite generally, explanations are based on the idea that the formation of a new solid phase at the expense of another one proceeds only at the interface between both phases.

Thus such an interface must first exist and once it is created it acts as a "catalyst" to spread the reaction. The interface grows and with it the rate. But sooner or later, the interface will start shrinking because of the progress of reaction and an S-shaped curve is obtained. The problem is then twofold. First we must describe how and at what rate the original reaction centers that provide the necessary interface are created. This is the problem of *nucleation*. Then we must understand at what rate the interface will advance through the unreacted solid phase. This is the problem of *growth*. It is doubtful that a single theory can account for all possibilities of nucleation and growth in all possible situations of interest. Furthermore, structural imperfections in solid structures are expected to affect profoundly rates of reactions involving these solids. These imperfections are of many types and can rarely be duplicated for a given compound because they depend, in a way difficult to control, on its previous history. Thus it has been known since Faraday that well formed crystals of hydrated salts will not effloresce over long periods of time in a dry atmosphere in which, according to thermodynamics, they should lose their water of crystallization. But these same crystals will effloresce very rapidly if their surface is marred by a scratch which provides the interface required for the new phase to spread.

These considerations are not limited to thermal decompositions of solids. They also apply to the reduction of many oxides by gaseous hydrogen or carbon monoxide. The rapid decomposition of solid explosive materials can sometimes be described by a rate equation of the type:

$$r = k \exp(\varphi t) \tag{6.4.3}$$

with an induction period t_i related to the parameter φ by means of the relation:

$$\varphi t_i = \text{constant} \tag{6.3.15}$$

that was shown to be characteristic of branched-chain reactions. In the present case, branched chains are not the explanation of the phenomenon. But the similarity in rate expressions shows again the autocatalytic behavior of these reactions involving solids.

The present brief discussion of reactions involving solids was kept on purpose at a phenomenological level. This reflects the underdeveloped state of this chapter in the kinetics of chemical change. Rapid progress is now being made in securing the molecular foundations of the field in the kinetics of nucleation, crystal growth and phase change in solids. An account of recent developments does not belong in this introduction to the subject.

Problem 6.4.1

Find the dependence of t_{max}, the time at which the maximum rate is reached, on the parameters k and n. Find also the value of f corresponding to the maximum rate when n is equal to 2 or 3.

6.5 Inhibition

The phenomenon of inhibition is so widely spread and so important to the study and understanding of chemical change, that it has even been dignified with the name of *negative catalysis*. *Inhibitors* themselves are known under various more or less specialized names: poisons, retarders, scavengers, traps, antioxidants, etc.

The kinetic definition of an inhibitor is a substance which decreases the rate of a chemical reaction. This it can do by interacting with the active centers in the sequence of elementary steps. Inhibitors can be of two types: temporary or reversible and permanent or irreversible. In the reversible category belong the inhibitors which will free the active centers when removed from the reacting mixture under reaction conditions. Irreversible inhibitors on the other hand stay bound to active centers when attempts are made to remove them from the reactor: They have destroyed active centers irreversibly and if the reaction is catalytic, the catalyst can be revived only through a re-activation procedure which may require drastic measures or may even be impractical.

In the case of irreversible inhibition, if the reaction is catalytic, its rate will reach zero value when all the sites responsible for catalytic action have been destroyed by the inhibitor or poison. The addition of known successive doses of poison then provides a method of counting the number of active sites on the enzyme or solid catalyst, provided that each molecule of poison destroys a constant number of sites. But in the case of a chain reaction, if the irreversible inhibitor is present in the reacting mixture from the very start, the reaction will be severely retarded until the inhibitor has all reacted away with the active centers produced in the initiation step. Thus during an induction period, the inhibitor disappears at a rate which provides a very useful measure of the rate of initiation.

These ideas can be expressed quantitatively in a simple manner. Consider a chain reaction with rate of initiation r_i, propagated through active centers **X** which are destroyed in a normal termination step at a rate $r_t = k_t(\mathbf{X})$ but also by reaction with an inhibitor D at a rate $r_d = k_d(\mathbf{X})(\mathrm{D})$. At the steady state:

$$\frac{d(\mathbf{X})}{dt} = r_i - k_t(\mathbf{X}) - k_d(\mathbf{X})(\mathrm{D}) = 0 \tag{6.5.1}$$

Hence, the steady-state concentration of active centers is given by:

$$(\mathbf{X})^* = \frac{r_i}{k_t + k_d(\mathrm{D})} \tag{6.5.2}$$

The rate of the chain reaction is that of the propagation steps:

$$r = a_p(\mathbf{X})^* = \frac{a_p r_i}{k_t + k_d(\mathrm{D})} \tag{6.5.3}$$

If the rate of destruction of active centers by the inhibitor is sufficiently larger than the normal rate of termination, i.e., if $k_d(\mathrm{D}) \gg k_t$, (6.5.3) becomes

$$r = \frac{a_p r_i}{k_d(\mathrm{D})} \tag{6.5.4}$$

so that the rate of reaction is simply inversely proportional to the concentration of inhibitor D. Similarly, the rate of initiation is then simply equal to the rate of destruction of active centers by the inhibitor, i.e., the rate of disappearance of the inhibitor is equal to the rate of initiation:

$$k_d(\mathbf{X}^*)(\mathrm{D}) = -\frac{d(\mathrm{D})}{dt} = r_i \tag{6.5.5}$$

If the initiator is present in sufficient excess, so that its concentration hardly changes during the induction period t_i, defined as the time required to consume all the inhibitor, we see that the latter disappears at a constant rate:

$$(\mathrm{D}) = (\mathrm{D})_0 - r_i t \tag{6.5.6}$$

where $(\mathrm{D})_0$ is the initial concentration of inhibitor.

Also:

$$r_i t_i = (\mathrm{D})_0 \tag{6.5.7}$$

Consequently the rate of initiation can be determined from a measurement of the induction period corresponding to a given amount of inhibitor.

Frequently, the action of the inhibitor is not as strong as that just depicted because, in a chain reaction, the inhibitor D reacts with an active center **X** to produce an inert molecule P and another active center **Y**:

$$\mathbf{X} + \mathrm{D} \rightarrow \mathrm{P} + \mathbf{Y}$$

This new active center **Y** is not, however, a stable entity but is able to participate in the chain propagation at a rate a_p' smaller than that, a_p, corresponding to the normal propagation step with **X**. At the steady state, with a termination rate constant for **Y** denoted by k_t':

$$\frac{d(\mathbf{X})}{dt} = r_i - k_t(\mathbf{X})^* - k_d(\mathbf{X})^*(\mathrm{D}) + a_p'(\mathbf{Y})^* = 0 \tag{6.5.8}$$

$$\frac{d(\mathbf{Y})}{dt} = k_d(\mathbf{X})^*(\mathrm{D}) - k_t'(\mathbf{Y})^* - a_p'(\mathbf{Y})^* = 0 \tag{6.5.9}$$

Thus:

$$(\mathbf{Y})^* = \frac{k_d(\mathbf{X})^*(\mathrm{D})}{k_t' + a_p'}$$

The rate of the inhibited reaction is now:

$$r = \frac{r_i a_p}{k_t + \dfrac{k_d(\mathrm{D})}{1 + \left(\dfrac{a_p'}{k_t'}\right)}} \tag{6.5.10}$$

A comparison with (6.5.3) shows that the difference between a strong inhibitor and a retarder is that the rate constant for inhibition k_d in the first case is replaced by a smaller one $k_d/[1 + (a'_p/k'_t)]$.

In many reactions of both chain and catalytic types, one of the reaction products acts like an inhibitor. In these frequent cases of self-inhibition, the rate may go down to zero before the end of the reaction can be reached. A more exceptional situation corresponds to the case where one of the reactants is an inhibitor. It might be expected that the rate would exhibit autocatalytic behavior as the reaction proceeds.

Such is the case in certain reactions catalyzed by enzymes. The type of inhibition considered thus far can be called competitive inhibition; the inhibitor competes with the reactant (called substrate S in enzyme kinetics) for the same active centers. But there also exists a different kind of inhibition called *noncompetitive inhibition*. A noncompetitive inhibitor D is one that combines with the enzyme E at a site which is different from that which combines with the substrate S. The complex **ED** between enzyme and inhibitor is then still able to combine further with a substrate molecule but the tertiary complex **EDS** thus formed is unreactive. If the rate-determining step of the reaction is the decomposition of the complex between enzyme **E** and substrate *S*, the sequence with noncompetitive inhibition can be represented as:

$$\mathbf{E} + \mathrm{S} \rightleftarrows \mathbf{ES} \qquad (1)$$

$$\mathbf{E} + \mathrm{D} \rightleftarrows \mathbf{ED} \qquad (2)$$

$$\mathbf{ED} + \mathrm{S} \rightleftarrows \mathbf{EDS} \qquad (3)$$

$$\mathbf{ES} + \mathrm{D} \rightleftarrows \mathbf{EDS} \qquad (4)$$

$$\mathbf{ES} \rightarrow \mathbf{E} + \mathrm{P} \qquad (5)$$

It shall be assumed further that the equilibrium constants for steps (1) and (3) are identical. So are the equilibrium constants for steps (2) and (4). Then it can be shown that the rate of reaction is:

$$r = (L)\frac{k_5}{[1 + K_2(\mathrm{D})]\left[1 + \dfrac{1}{K_1(\mathrm{S})}\right]} \qquad (6.5.11)$$

If it happens that the inhibitor D is identical to the substrate S, the rate equation (6.5.11) becomes:

$$r = (L)\frac{k_5}{[1 + K_2(\mathrm{S})]\left[1 + \dfrac{1}{K_1(\mathrm{S})}\right]} \qquad (6.5.12)$$

At sufficiently high values of (S), this reduces to

$$r = (L)\frac{k_5}{K_2(\mathrm{S})} \tag{6.5.13}$$

At sufficiently low values of (S), (6.5.12) becomes:

$$r = (L)k_5K_1(\mathrm{S}) \tag{6.5.14}$$

Thus, from (6.5.13), the rate increases with increasing extent of reaction but from (6.5.14), it behaves normally, i.e., it decreases with extent of reaction.

Consequently, at intermediate values of conversion, the rate will go through a maximum. At first glance, it might seem that the reaction is autocatalytic. In reality, it is inhibited by the reactant itself. This unusual result shows the narrow connection between catalysis, autocatalysis and "negative" catalysis; the kinetic patterns are dictated by the reactions with the active centers, their multiplication or their destruction.

Problem 6.5.1

Derive Eq. (6.5.11).

BIBLIOGRAPHY

6.1 Other classical cases of autocatalysis as well as many types of "formal kinetics" are discussed in a monumental work: *Cinétique Chimique Appliquée* by Jungers, *et al.*, Chapter IV, Technip, Paris, 1958.

6.2 A thorough discussion of branching and degenerate branching can be found in
6.3 Semenov's *Some Problems in Chemical Kinetics and Reactivity*, translated by M. Boudart, Vol. II, Princeton University Press, Princeton, N.J., 1959.

6.4 A very good review by Tompkins and Young of the kinetics of reactions involving solids is found in *Ann. Reports Chem. Soc.* (London) **50**, 70 (1954). More detailed references dealing with the various types of rate equations can be found there. The original paper by G.N. Lewis is in *Z. Physik. Chem.* **52**, 310 (1905).

6.5 Many interesting examples of "negative catalysis" are found in P.G. Ashmore's *Catalysis and Inhibition of Chemical Reactions*, Butterworths, London, 1963. This book may be considered as a successor to the classic *Catalysis from the Standpoint of Chemical Kinetics*, by Georg-Maria Schwab, translated by Hugh S. Taylor and R. Spence, D. van Nostrand Company, Inc., New York, 1937.

Irreducible Transport Phenomena in Chemical Kinetics

7

Transfer of heat and mass is essential to the treatment of rates of reaction in many situations that can be grouped in two categories. In the first category, transport phenomena and, in particular, the flow characteristics that control them are intimately connected with a *scale* factor. For instance, heat removal in a fixed bed of solid particles catalyzing an exothermic reaction between fluid reactants may not be a problem if the bed is small but may become quite serious in a large reactor. On the contrary, in the case of such a bed, transport of reactants to the external surface of the particles may be more than adequate in a large column with high velocities of flow but it may be limiting in a laboratory-size reactor with small rates of flow.

Frequently the coupling between physical and chemical factors can be said to be irreducible; it is essential to the phenomenon itself. Thus, methane burning in the flame of a Bunsen burner reacts at a rate which is determined by physical and chemical processes that could be uncoupled only by destroying the flame.

However, in this chapter we shall not treat the category of such problems where flow is an essential component of the situation. Rather we shall look at a number of effects due to the irreducible *coupling between molecular diffusion of heat or mass and the chemical reaction.*

7.1 The Gel and Cage Effects

The rate of rapid reactions in the liquid phase can be controlled by diffusion. Among such reactions are the termination steps of some chain reactions. In particular, the viscosity of a solution in which polymerization takes place increases rapidly as the reaction proceeds. Also the growing polymer chains that recombine in the termination step become very large and diffuse with greater difficulty in the viscous medium. Because the rate constants of termination steps are usually very large, these steps become controlled by diffusion. As a result, the rate of the termination step goes down and the steady-state concentration of active centers goes up. The propagation steps are hardly affected by the change in viscosity because they are considerably slower and involve only one large molecule, the other being a monomer. Therefore, as the steady-state concentration of active centers increases, the rate of polymerization also becomes higher as the extent of reaction increases. This is another example of an exception to Rule I of Chapter 1 which states that the isothermal rate of reaction decreases with advancement of reaction. The phenomenon, called the *gel effect*, is due to a change in properties of the reaction medium and to the fact that one of the steps in the chain becomes diffusion controlled.

Limiting conditions under which a homogeneous bimolecular step between two species A and B can become controlled by diffusion are easily derived, at least in an elementary fashion which does not claim to be rigorous. Early theoretical attention was given to these phenomena by Smoluchowski (1917), Onsager (1934), and more recently by Debye (1942). Let us focus our attention on the molecules of the A type. Each one can be conceived as surrounded by a spherical surface of radius r through which B molecules diffuse. If r is large enough, say infinity, the concentration of B molecules at that radius is equal to $(B)_\infty$, the average concentration of B molecules in solution. But if the reaction between A and B is very fast, the concentration of B molecules around each A molecule becomes depleted: $(B)_r < (B)_\infty$. A diffusion flux sets in to replenish the B molecules that are consumed in the reaction. This flux must be equal to the reaction rate.

The latter would normally be given by:

$$r = k(A)(B) \tag{7.1.1}$$

Since we focus our attention on the A molecules, we consider the reaction rate r', per unit concentration of A molecules:

$$r' = k(\mathrm{B}) \tag{7.1.2}$$

This must be equal to the diffusion flux:

$$r' = DS\frac{d(\mathrm{B})}{dr} \tag{7.1.3}$$

where D is the diffusivity and S the area of a sphere of radius r. Therefore:

$$r' = D4\pi r^2\frac{d(\mathrm{B})}{dr} \tag{7.1.4}$$

This diffusion equation can be integrated between $r = \infty$ and $r = \rho$, where ρ is the average radius of A and B molecules. The corresponding values of (B) are: $(\mathrm{B})_\infty$ and $(\mathrm{B})_\rho$. The result of the integration of (7.1.4) is then:

$$(\mathrm{B})_\infty - (\mathrm{B})_\rho = \frac{r'}{4\pi D\rho} \tag{7.1.5}$$

But r' is equal to the actual rate of reaction taking place:

$$r' = k(\mathrm{B})_\rho \tag{7.1.6}$$

If there did not exist any diffusional limitation, the rate would have an ideal value:

$$r'_{\text{ideal}} = k(\mathrm{B})_\infty \tag{7.1.7}$$

The ratio of actual to ideal rate may be called an *efficiency factor* η which measures the diffusional limitation:

$$\eta = \frac{r'}{r'_{\text{ideal}}} = \frac{(\mathrm{B})_\rho}{(\mathrm{B})_\infty} \tag{7.1.8}$$

Eliminating $(\mathrm{B})_\rho$ and $(\mathrm{B})_\infty$ between Eqs. (7.1.5), (7.1.6), and (7.1.8) we get:

$$\boxed{\eta = \frac{1}{1 + \dfrac{k}{4\pi\rho D}}} \tag{7.1.9}$$

The actual rate of reaction will be:

$$r' = \eta k(\mathrm{B})_{\infty}$$

From (7.1.9), we see that no diffusional limitation will be encountered, i.e., the efficiency η approaches unity when

$$k \ll 4\pi\rho D \tag{7.1.10}$$

On the contrary, the rate of reaction is controlled by diffusion if

$$k \gg 4\pi\rho D \tag{7.1.11}$$

As an illustration of these limiting conditions, consider the rate of termination by mutual combination of two polymeric active centers of high molecular weight. Suppose that these active centers can be considered to have a spherical shape of radius $\rho = 1000$ Å. The diffusivity of these species can be obtained, in order of magnitude from the Stokes-Einstein relation:

$$D = \frac{kT}{6\pi\rho\mu} \tag{7.1.12}$$

where μ is the viscosity of the solution, say one poise. It will be assumed that recombination takes place without any activation barrier but with a probability factor equal to 10^{-2}. Then $k \cong 10^{-2} \times 10^{-10} = 10^{-12}$ cm³/sec. The critical condition becomes, at 300°K:

$$10^{-12} = k \gg \frac{4\pi kT}{6\pi\mu} = \tfrac{2}{3}(1.38) \times 10^{-16} \times 300 \cong 3 \times 10^{-14}\ \mathrm{cm^3/sec.}$$

Consequently, this would be a clear case of a diffusion-controlled reaction, as expressed by the inequality (7.1.11).

In water at 25°C, with a viscosity of one centipoise, the critical condition would be more severe: 3×10^{-12} cm³/sec. This condition is clearly violated by the experimentally well-established rate constant for the reaction

$$\mathrm{H^+ + OH^- \rightarrow H_2O}$$

This rate constant is now known to be equal to 2.1×10^{-10} cm³/sec. This extremely high value of a rate constant of a diffusion-limited step in solution is made possible by the mobility of a proton which can shift along a hydrogen bonded chain after the ions have diffused to within 6 to 8 Å of each other: $\mathrm{H_2OH^+ \ldots HOH \ldots HOH \ldots OH^- \rightarrow H_2O \ldots HOH \ldots HOH \ldots HOH}$

Conversely, diffusion is also a factor when a molecular species dissociates into fragments either thermally or as a result of absorption of energy due to light or ionizing radiation.

In a dense gas or liquid phase, the fragments may not be able to diffuse away from each other through the surrounding molecules before they recombine; it is as if they were prisoners in a cage of solvent molecules. Hence the name *cage effect* for this phenomenon first discussed by Rabinowitch (1934).

Consider, for instance, the photolysis of azomethane:

$$CH_3NNCH_3 \leadsto 2\,CH_3 + N_2$$

The methyl radicals CH_3 produced from a given molecule of azomethane may well recombine to form ethane:

$$CH_3 + CH_3 \rightarrow C_2H_6$$

before they have a chance to diffuse away from each other. Indeed when an equimolar mixture of azomethane and perdeuteroazomethane CD_3NNCD_3 is photolyzed in isooctane solution, the ethane molecules that are formed are exclusively C_2H_6 and C_2D_6. No mixed ethane CH_3CD_3 can be found. By contrast, when the same experiment is done in the gas phase, twice as many mixed molecules CH_3CD_3 are found as either one of the unmixed species C_2H_6 or C_2D_6. This is a striking demonstration of the cage effect in solution and of its absence in the gas phase where the free radicals CH_3 and CD_3 recombine at random.

Cage effects tend to decrease the efficiency of utilization of free radicals produced for the initiation of chain reactions. A quantitative treatment appears in sight but will not be discussed here.

Problem 7.1.1

In a given liquid-phase polymerization process, the addition of a chain transfer agent to the reaction mixture is found to suppress the "peaking" of the rate (gel effect) that occurs without this agent. Rationalize this observation.

7.2 *The Wall Effect*

The rate of a homogeneous reaction, r, should not depend on the scale of the reactor in which the reaction is carried out. When this happens, a diffusional process must be suspected. In the preceding section, it was seen how diffusional limitations could change the rate of liquid-phase chain reactions by affecting the rate of termination. In the gas phase at low pressures, homogeneous termination is frequently accompanied by termination at the walls of the reactor. This heterogeneous mode of destruction of active centers may compete successfully with homogeneous destruction or even become the main termination step. Solid surfaces of almost any material are able to capture free radicals (active centers) with a high probability per collision. Destruction of the active center follows capture and only an inert product leaves the wall. *Wall effects* are very general in gaseous chain reactions. Because the catalytic activity of the wall is sensitive to impurities and difficult to reproduce or control, wall effects are a nuisance in the study of chain reactions. In practice, the success or failure of a process consisting of a homogeneous chain reaction occasionally rests on the catalytic activity or lack of activity of the material lining the reactor.

Because the rate of wall termination is frequently high, it is often controlled by the diffusion of active centers to the internal surface of the reactor and this is an irreducible situation which can well be changed by altering flow conditions but cannot be eliminated altogether.

It will now be shown how the rate of a chain reaction depends on pressure, reactor size and reactor shape when the transport of active centers (chain carriers) **X** to the wall is purely diffusive. It is assumed that homogeneous termination can be neglected. Then there are no homogeneous sinks of active centers. There is only a homogeneous source corresponding to the usual rate of initiation r_i. In a reactor consisting of two parallel plates sufficiently large so that end effects are immaterial and at a distance L from each other, the differential equation, the solution of which gives $(\mathbf{X})^*$ as a function of the coordinate z measuring distance between plates, is written as follows:

$$D\frac{d^2(\mathbf{X})^*}{dz^2} + r_i = 0 \tag{7.2.1}$$

where D is the diffusivity of **X**. The way to derive such an equation will be explained in the next section (see 7.3.7 to 7.3.11). The concentration of active centers is maximum at the center ($z = 0$) and equal to zero at the wall ($z = L/2$). Therefore the boundary conditions are:

$$\frac{d(\mathbf{X})^*}{dz} = 0 \quad \text{at} \quad z = 0$$

$$(\mathbf{X})^* = 0 \quad \text{at} \quad z = \frac{L}{2}$$

The solution of (7.2.1) is then:

$$(\mathbf{X})^* = \frac{r_i}{2D}\left(\frac{L^2}{4} - z^2\right) \tag{7.2.2}$$

The extensive rate of reaction in a volume of thickness dz and unit area is:

$$dR = a_p(\mathbf{X})^* dz$$

The extensive rate of reaction in the reactor, also assumed to be of unit area, is then:

$$R = 2\int_0^{L/2} a_p(\mathbf{X})^* dz \qquad (7.2.3)$$

The average specific rate of reaction per unit volume of reactor is R divided by the reactor volume:

$$\bar{r} = \frac{2}{L}\int_0^{L/2} a_p(\mathbf{X})^* dz \qquad (7.2.4)$$

Substituting the value of $(\mathbf{X})^*$ given by (7.2.2) and integrating (7.2.4), we get

$$\bar{r} = r_i a_p \frac{L^2}{12D} \qquad (7.2.5)$$

The interpretation of this result is clear if use is made of Einstein's equation relating D and the mean square distance $\overline{L^2}$ covered by a diffusing particle in a time t:

$$\overline{L^2} = 2Dt \qquad (7.2.6)$$

In our case, this time t is the average lifetime of an active center in the termination step, as previously defined:

$$t = \frac{1}{a_t} \qquad (7.2.7)$$

Also

$$\overline{L^2} = L^2 \qquad (7.2.8)$$

Substituting (7.2.6), (7.2.7), and (7.2.8) in (7.2.5), we get:

$$\boxed{\bar{r} = r_i \frac{a_p}{6a_t}} \qquad (7.2.9)$$

where Γ is a numerical coefficient of the order of unity which is determined by the symmetry of the problem, i.e., the shape of the reactor. In this case $\Gamma = 6$. The result expressed in (7.2.9) says that *the average rate of reaction is equal, as usual, to the rate of initiation times the kinetic chain length* (see p. 72). *But the latter is modified by a numerical factor* Γ *multiplying* a_t.

It is interesting that the same result could have been obtained immediately, without solving the differential equation, by writing the customary equation determining the concentration of active centers at the steady state, but multiplying a_t by Γ:

$$\frac{d(\mathbf{X})}{dt} = r_i - \Gamma a_t(\mathbf{X})^* = 0 \tag{7.2.10}$$

The difference is that $(\mathbf{X})^*$ coming from this equation would have only a space average value and of course there is no way to determine the numerical value of Γ in such a fashion. But this simplification is of great practical value if the correct differential equation of the problem is too complicated. Then, to find the dependence of the rate of reaction on pressure and reactor size, it is only necessary to solve the algebraic equation (7.2.10) and remember at the end that, following the Einstein relation (7.2.6):

$$a_t = \frac{2D}{L^2} \tag{7.2.11}$$

For instance, in the case just treated $\bar{r}$ is found to be proportional to the square of the characteristic size of the reactor. It is also proportional to the square of the total pressure p if it is assumed that the rate of initiation is independent of pressure. Indeed, a_p is proportional to p, and D, the diffusion coefficient of the active center, is inversely proportional to p.

The problem can be treated in a similar way for the case of a cylindrical or a spherical reactor. Only, respectively, the differential equation (7.2.1) would be replaced by:

$$D\frac{1}{z}\frac{d}{dz}\left(z\frac{d(\mathbf{X})}{dz}\right) + r_i = 0 \tag{7.2.12}$$

$$D\frac{1}{z^2}\frac{d}{dz}\left(z^2\frac{d(\mathbf{X})}{dz}\right) + r_i = 0 \tag{7.2.13}$$

where z is the radial coordinate. Solution gives the corresponding values of Γ which are only slightly different from that for the plane reactor.

Naturally, the fact that termination takes place on the wall does not necessarily mean that this step is controlled by diffusion. This will not be the case if the probability of capture by the wall is small, the pressure is low and the size of the reactor is not too large. For diffusion to be rate-controlling, the rate of reaction per unit surface area of wall must be much larger than the maximum flux due to diffusion.

If active centers that strike the wall at a frequency $v/4$ (see Table 2.5.1, p. 47) are captured at a probability γ per collision, the first quantity is equal to

$$\frac{v}{4}\gamma\,(\mathbf{X})$$

The second quantity is of the order of

$$D\frac{(\mathbf{X})}{L}$$

Thus, diffusion will control wall termination if:

$$\frac{v}{4}\gamma \gg \frac{D}{L} \tag{7.2.14}$$

But from gas kinetic theory:

$$D = \tfrac{1}{3}v\lambda \tag{7.2.15}$$

where λ is the mean free path.

The condition becomes:

$$\boxed{\gamma \gg \frac{\lambda}{L}} \tag{7.2.16}$$

Thus diffusion to the wall will control the rate of termination if the probability of destruction of active centers per collision with the wall is larger than the ratio of two characteristic dimensions: the mean free path and the size of the reactor.

Problem 7.2.1

Find the values of Γ for a cylindrical or a spherical reactor of diameter d.

Problem 7.2.2

Consider again the stoichiometric hydrogen-oxygen mixture at 800°K treated in Problem 6.2.1. At very low pressure, termination is due only to destruction at the wall of the hydrogen atoms:

$$\mathbf{H} + \text{wall} \quad \rightarrow \quad \text{inactive molecule}$$

This proceeds at a rate:

$$r_t = \frac{v}{4}\gamma(\mathbf{H})$$

Find the so-called first explosion limit in a cylindrical reactor 1 cm in diameter. This is the critical pressure at which the net branching factor will be equal to zero. Assume a value $\gamma = 10^{-3}$ for the probability of capture of hydrogen atoms at the wall. Assume that diffusion is not controlling and from the calculated explosion limit, verify the validity of this last assumption. In setting up the critical condition for explosion, reflect upon the fact that r_t is now the rate of a surface reaction while the branching still occurs homogeneously throughout the volume of the reactor.

7.3 *The Penetration Effect*

Consider a solid, say nickel, reacting with a gas, say oxygen, to form another coherent, nonporous new solid phase, e.g., nickel oxide. For reaction to take place, one of the reactants (either nickel or oxygen ions) must diffuse through the film consisting of the new solid phase. If L is the thickness of the film at time t, the rate of diffusion, i.e., the rate of reaction r, will be inversely proportional to the film thickness and it will also be proportional to dL/dt. Therefore:

$$\frac{dL}{dt} = \frac{C}{L} \tag{7.3.1}$$

where C is a proportionality constant. Integration of (7.3.1) gives:

$$\boxed{L = \sqrt{2Ct}} \tag{7.3.2}$$

This is the *parabolic law* found in the oxidation of certain metals.

Frequently the effect of penetration is somewhat different. A very simple experiment illustrates the nature of the problem. From a single block of porous graphite, let us cut out two small cylinders L and S. The large cylinder L has a height and diameter equal to 12 mm. The small one S has a height and diameter equal to 6 mm. Let us heat these cylinders in oxygen at a given temperature and measure the ratio ρ of losses of weight of L and S in a given time. At low temperatures, it is found (Fig. 7.3.1) that $\rho = 8$. As the

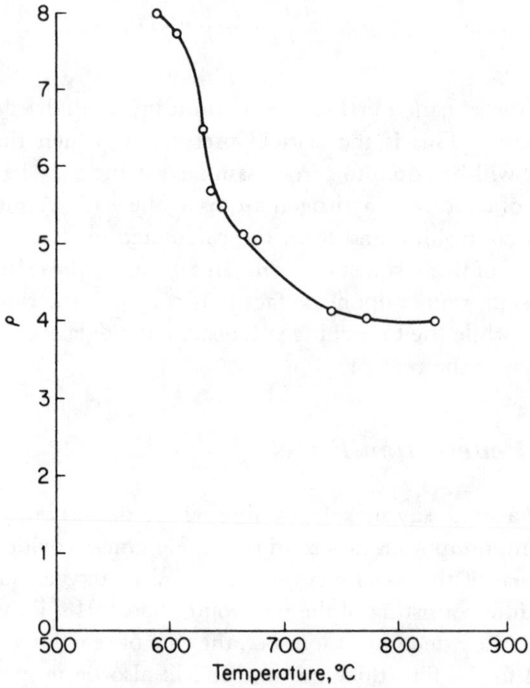

Fig. 7.3.1 The effect of penetration. Relative oxidation rates vs temperature for two graphite cylinders of linear dimensions in the ratio of two. [X. Duval, *J. Chim. Phys.*, **58,** 5 (1961)]

experiment is repeated at increasingly higher temperatures, it is found that around 600°C, the ratio ρ starts to fall. When the temperature has reached about 800°C, ρ is equal to 4 and this value stays constant at still higher temperatures. What happens between 600 and 800°C and why is ρ equal to 8 at low temperatrues and only 4 at high temperatures?

This result is explained as follows. The linear dimensions of the two cylinders are in the ratio 2 to 1. Their external surface areas are in the ratio 4 to 1. Their volumes or masses are in the ratio 8 to 1. It follows that everything happens as if rates of oxidation were proportional to volumes at low temperatures but to surfaces at high temperatures. This can be explained if it is assumed that the internal surface area proportional to the mass of the sample is accessible to oxygen at low temperatures. At higher temperatures, the rate of oxidation becomes too fast and an adequate supply of oxygen cannot diffuse fast enough into the internal network of pores. Finally, at sufficiently high temperatures, the reaction rate is so fast that none of the internal surface plays any role and only the external surface of the sample is reacting away.

This is a general situation: When reactants, gaseous or liquid, have to diffuse from one phase to another phase which may be a liquid or a porous solid with which these reactants react or in which they react catalytically, this *effect of penetration* will be more or less marked depending on the relative rates of reaction and diffusion. Since many solid catalysts consist of porous grains or pellets with a large internal surface area, this phenomenon will be of importance in all catalytic processes using such solids.

Consider in general a horizontal layer of reactive or catalytic phase of thickness L and unit surface area. Reactant A at a concentration $(A)_0$ enters this phase at the upper interface. If this external concentration could be maintained by diffusion throughout the layer, the extensive rate of reaction which we may call ideal would be:

$$R_{\text{ideal}} = k(A)_0 L \tag{7.3.3}$$

It is assumed that A reacts in a first-order reaction with rate constant k.

But since A reacts as it penetrates, its concentration will decrease across the layer. The effective average length of penetration of A will be given by Einstein's relation (see p. 150):

$$\overline{L}^2 = 2Dt \tag{7.2.6}$$

where t is the average lifetime of reacting A, namely $1/k$, and D is the diffusivity of A′.

Therefore the real rate of reaction is only:

$$R_{\text{real}} = k(A)_0(\overline{L}^2)^{1/2} = k(A)_0\left(\frac{2D}{k}\right)^{1/2} \tag{7.3.4}$$

The ratio of real and ideal rates is an efficiency of utilization of the reactive or catalytic layer:

$$\boxed{\eta = \frac{R_{\text{real}}}{R_{\text{ideal}}} = \frac{1}{L}\sqrt{\frac{2D}{k}}} \tag{7.3.5}$$

It is evident that the efficiency will be small if the layer is thick, the diffusivity low or the rate constant large.

Let us refine this crude argument by investigating diffusion and reaction in a long narrow cylinder, the walls of which are covered with catalytic material. This reactor can serve as an idealized model of a pore in a catalyst pellet. Let the length of reactor be L, the length coordinate z ($z = 0$ at the reactor entrance), the radius of the reactor ρ. Reactant A penetrates into the reactor by diffusion only. Its concentration at $z = 0$ is $(\mathrm{A})_0$. Its diffusivity is D. Reaction on the walls proceeds at a rate:

$$r = k(\mathrm{A}) \tag{7.3.6}$$

corresponding to a first-order reaction.

Consider a slice of reactor of length dz. The number of molecules of A diffusing into this slice is:

$$-D\pi\rho^2 \frac{d(\mathrm{A})}{dz} \tag{7.3.7}$$

The number of molecules of A diffusing out of this slice is:

$$-D\pi\rho^2 \frac{d}{dz}\left[(\mathrm{A}) + \frac{d(\mathrm{A})}{dz}\,dz\right] \tag{7.3.8}$$

The number of molecules of A reacting away in this slice is:

$$r2\pi\rho\, dz = 2\pi\rho\, k(\mathrm{A})\, dz \tag{7.3.9}$$

The material balance states that

$$(7.3.7) - (7.3.8) = (7.3.9)$$

The equation of the problem is then:

$$D\pi\rho^2 \frac{d^2(\mathrm{A})}{dz^2} = 2\pi\rho k(\mathrm{A}) \tag{7.3.10}$$

Hence:

$$\frac{d^2(\mathrm{A})}{dz^2} = \frac{2k}{\rho D}(\mathrm{A}) \tag{7.3.11}$$

To integrate (7.3.11), we rewrite it as follows:

$$\frac{d^2(\mathrm{A})}{dz^2}\frac{d(\mathrm{A})}{dz}\frac{dz}{d(\mathrm{A})} = \frac{2k}{\rho D}(\mathrm{A})$$

$$\tfrac{1}{2}d\left[\frac{d(\mathrm{A})}{dz}\right]^2 = \frac{2k}{\rho D}(\mathrm{A})\,d(\mathrm{A}) \tag{7.3.12}$$

Integration gives:

$$\left[\frac{d(\mathrm{A})}{dz}\right]^2_{z=L} - \left[\frac{d(\mathrm{A})}{dz}\right]^2_{z=0} = \frac{4k}{\rho D}\int_{(\mathrm{A})_0}^{(\mathrm{A})} (\mathrm{A})\,d(\mathrm{A}) \tag{7.3.13}$$

It will now be assumed that the penetration effect is severe enough so that $(\mathrm{A})_L = 0$ at $z = L$. Thus the gradient of concentration also vanishes at $z = L$:

$$\left[\frac{d(\mathrm{A})}{dz}\right]_{z=L} = 0$$

It follows that:

$$\left[\frac{d(\mathrm{A})}{dz}\right]_{z=0} = \pm\left[\frac{4k}{\rho D}\frac{(\mathrm{A})_0{}^2}{2}\right]^{1/2} \tag{7.3.14}$$

The minus sign is to be chosen because the gradient of concentration is negative.

The extensive rate of reaction in the reactor is equal to the entering flux:

$$R_{\text{real}} = -D\pi\rho^2\left[\frac{d(\mathrm{A})}{dz}\right]_{z=0} = \pi\rho^2 D\left(\frac{2k}{\rho D}\right)^{1/2}(\mathrm{A})_0 \tag{7.3.15}$$

If, on the other hand, diffusion had been adequate so that the initial concentration $(\mathrm{A})_0$ could have been maintained throughout the reactor, the ideal extensive reaction rate would have been:

$$R_{\text{ideal}} = 2\pi\rho Lk(\mathrm{A})_0 \tag{7.3.16}$$

The ratio of actual to ideal rates is again a measure of the efficiency of utilization of the available internal catalytic surface:

$$\boxed{\frac{R_{\text{real}}}{R_{\text{ideal}}} = \eta = \frac{1}{L}\sqrt{\frac{\rho D}{2k}}} \tag{7.3.17}$$

The inverse of the efficiency η is a dimensionless group frequently called the Thiele modulus h. This important problem was treated independently in 1939 by Thiele and Zeldovich. By solving the differential equation (7.3.13) without the assumption of a severe penetration effect (namely (A) $= 0$ at $z = L$), it can be shown that indeed $\eta = 1/h$ for values of h larger than five. Severe penetration then means $h \geq 5$. This case is also referred to as that of the *internal diffusion regime.*

The complete solution valid for smaller values of h does not need to be considered in an elementary presentation. Indeed, in order to apply this treatment to catalyst particles or pellets, it is necessary to have a model of the porous network. In order to apply the results, it is necessary to know the effective diffusivity in the porous material under conditions of reactions. This information is rarely available. Furthermore, solutions are not available for the types of rate functions encountered in practice.

Therefore, within the scope of this treatment, only the critical condition for diffusional limitation will be given attention. An order of magnitude value of the effective diffusivity in a porous medium is readily available when the diffusing species is a gas and when the mean free path of the gas is large in comparison to the average diameter δ of the pores. Then, the analog of the approximate gas kinetic formula:

$$D = \tfrac{1}{3}v\lambda \qquad (7.2.15)$$

is for so-called Knudsen diffusion in the porous medium (i.e. $\lambda \gg \delta$)

$$D = \tfrac{1}{3}v\delta \qquad (7.3.18)$$

where the characteristic length λ for diffusion in the gas has been replaced by δ, the average diameter of the pores.

For a first-order surface reaction, the rate constant may be expressed as the product of the collision frequency $(v/4)$ and the probability γ of reaction per collision.

Then the condition for a severe penetration effect or for the regime of internal diffusion:

$$L\left(\frac{2k}{\rho D}\right)^{1/2} \gg 1 \qquad (7.3.19)$$

becomes, if we note that $\delta = 2\rho$:

$$\boxed{\gamma^{1/2} \gg \frac{\delta}{L}} \qquad (7.3.20)$$

This condition must be compared to (7.2.16) for the case of a diffusion-controlled wall termination:

$$\gamma \gg \frac{\lambda}{L} \tag{7.2.16}$$

Except for the fact that in the case of a porous medium the probability of reaction per collision is raised to the one-half power, these two conditions may be considered as identical: The characteristic size L of the homogeneous reactor where the chain reaction proceeds is replaced by the characteristic thickness L of the porous layer where the catalytic reaction proceeds. The characteristic length for molecular motion in the gas phase, namely the mean free path λ, is replaced by the characteristic length for molecular motion in the porous medium, the average diameter of the pores δ. Thus kinetic phenomena that seem at first glance to be very different, are actually governed by very similar laws.

Problem 7.3.1

By arguments similar to (7.3.8), (7.3.9), (7.3.10), and (7.3.11), derive the differential equation (7.2.1).

Problem 7.3.2

Justify the exponential decrease in reactant concentration illustrated in Fig. 7.3.2.

Problem 7.3.3

Show that both expressions for η, (7.3.4) and (7.3.16), are identical within a numerical factor of the order of unity. Note that both rate constants k have a different meaning but they are related to each other.

Problem 7.3.4

What is the value of the apparent activation energy of a reaction the rate of which is measured in a regime controlled by diffusion through the pores of a catalyst pellet, if the true activation energy of the first-order reaction is equal to E?

Problem 7.3.5

What is the apparent order of a reaction that is known to be second-order if its rate is measured in a regime controlled by diffusion through the pores of a catalyst pellet?

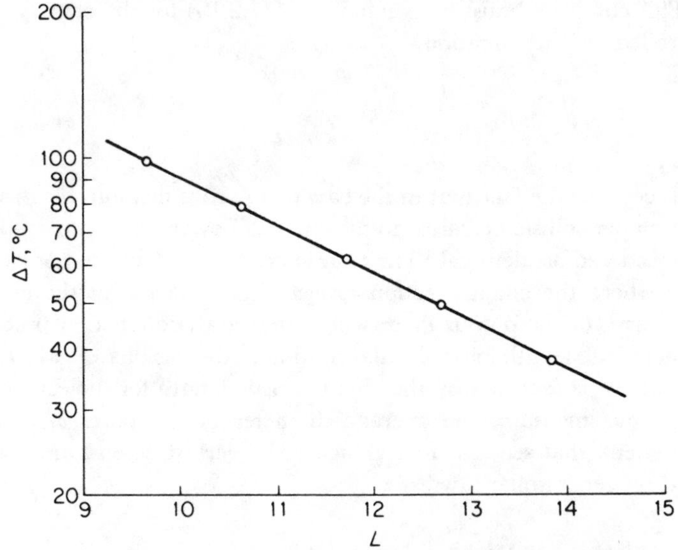

Fig. 7.3.2 Exponential decay of hydrogen atoms in a tube with catalytic walls. These data were obtained by traveling a thermocouple probe axially down a closed Pyrex tube. At one end of the tube, a discharge generates H atoms. The atoms diffuse down the tube and recombine catalytically on the walls following a first-order reaction. The probe is covered with a very active catalyst for the exothermic recombination. As a result, its temperature is higher than that of the walls by an amount ΔT, which is proportional to the concentration of atoms at the location L of the probe. This dimensionless distance L is the distance down the tube divided by the tube radius. [See K. Tsu and M. Boudart, *Actes 2e Congrès Intern. Catalyse*, Vol. I, p. 593, Technip, Paris (1961).]

7.4 *Thermal Explosions, Ignition, Stability*

It has been shown in the preceding chapter how the rate of an autocatalytic reaction could accelerate in an isothermal system and lead in principle to an isothermal explosion. We wish to treat in this section the case of *thermal explosions* which can happen in a reactor where an exothermic reaction takes place if the heat of reaction cannot be removed rapidly enough, i.e., if a heat transfer phenomenon becomes limiting.

Consider (Fig. 7.4.1) a plot of rate of heat release (Curve 1) versus temperature. The rate of heat release is directly proportional to the rate of reaction and the latter increases exponentially with temperature. On the same diagram, let us plot the rate of heat removal (Curve 2) which will be approximately a straight line (Newton's Law of Cooling) intersecting the axis of temperatures at a value T_0 equal to the wall temperature. Curve 2 will frequently intersect the exponential curve and then it will do so at two points corresponding to a lower temperature T_1 and to a higher temperature T_2. At these points, the reaction can proceed at a thermal steady state, since the rate of heat release is equal to the rate of heat removal.

However, it is easily seen that the state corresponding to the higher temperature is *unstable* while that corresponding to the lower temperature is *stable*. In order to examine this question of stability, the state of interest is submitted to a small perturbation on either side of the steady state. The system is stable if it tends to return to the original state. It is unstable if it tends to move away

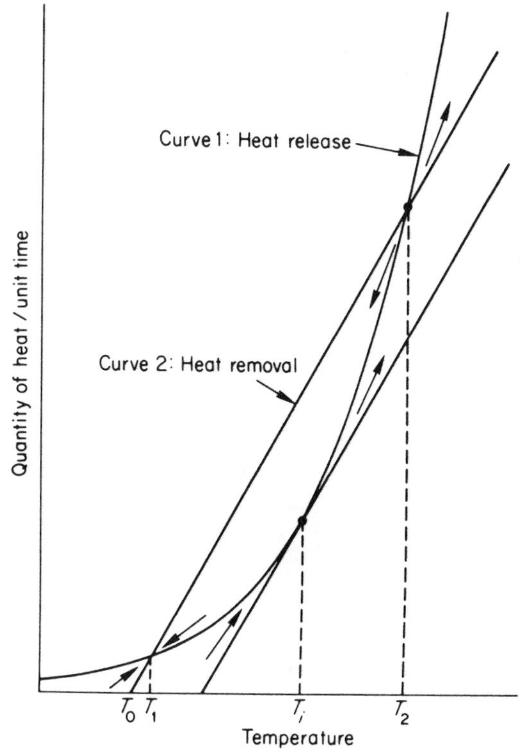

Fig. 7.4.1 Stable and unstable regimes of operation for an exothermic reaction.

from it. In our case, the direction of motion will be dictated by the comparative values of the rates of heat release and of heat removal away from the steady state. The lower temperature steady state is a stable state because at $T_1 + \Delta T$ removal of heat is faster than its production so that the temperature drops back to T_1, while at $T_1 - \Delta T$ the converse is true and the temperature moves up to T_1. This illustrates how the coupling between chemical kinetics and heat transfer determines the thermal steady-state evolution of a reacting chemical system and also the stability or instability of operation.

If now the wall temperature T_0 is raised, the line of the rate of heat removal will move parallel to itself and ultimately it becomes tangent to the exponential curve (1). The point at which both curves are tangent to each other determines a temperature T_i that can be called an *ignition temperature*. Indeed this is the highest possible temperature for a steady-state process. Beyond that temperature, Curves 1 and 2 do not meet each other, the process is always unstable since the rate of heat release always exceeds that of heat removal. The rate will accelerate without check as the temperature of the system rises and a thermal explosion occurs.

Normally, these qualitative considerations can be refined by considering

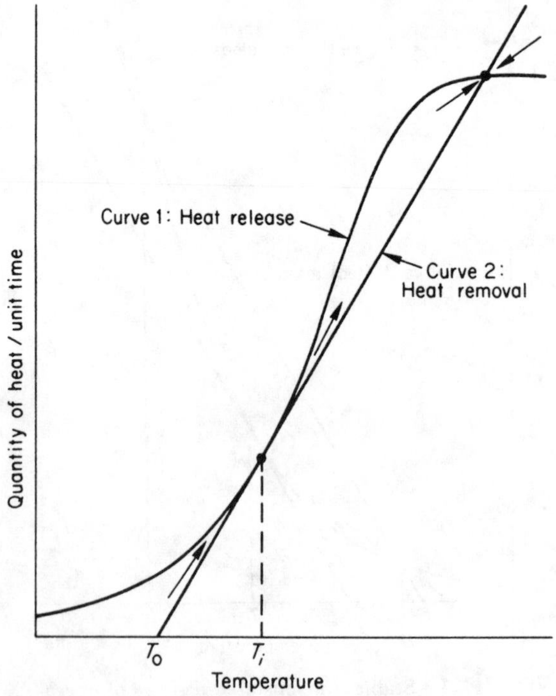

Fig. 7.4.2a Ignition on a catalytic surface.

for instance the temperature distribution inside the reactor. Similar considerations apply to many other situations. For instance, if an exothermic reaction is catalyzed at the surface of a solid, the curve of rate of heat release does not keep increasing exponentially with surface temperature. This is because the rate of reaction at the surface, as it becomes larger and larger, becomes limited by the rate of diffusion of reactants to the surface and the latter increases only slightly with temperature. Thus the curve of the rate of heat release is S-shaped (Fig. 7.4.2a). Above a certain "ignition" temperature, determined as above, the surface temperature jumps to a rather high value determined by the intersection of the curve of rate of heat removal with the upper branch of the curve of rate of heat release (Fig. 7.4.2b). This high *surface ignition* may impair or destroy the activity of the catalytic surface. The effect involves a coupling between rate of reaction and both transport phenomena of heat and mass.

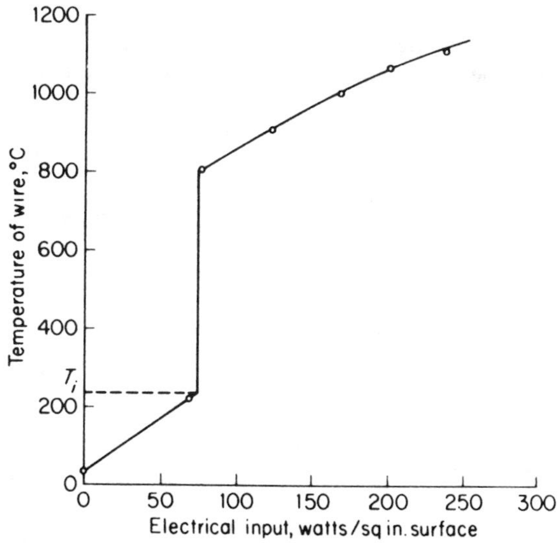

Fig. 7.4.2b Example of ignition on a catalytic surface. Temperature of platinum wire (0.001 in. dia.) vs power input in the presence of 3.5% hydrogen and 96.5 % air. [W. Davies *Phil. Mag.*, **17**, 235 (1934)] During the catalytic reaction of hydrogen and oxygen on the platinum, when the wire temperature reaches T_i, it suddenly jumps to the higher ignition temperature as described in Fig. 7.4.2a.

BIBLIOGRAPHY

This chapter gives only a glimpse of a vast field surveyed by D. A. Frank-Kamenetskii in *Diffusion and Heat Exchange in Chemical Kinetics*, translated by N. Thon, Princeton University Press, Princeton, N.J., 1955. This book, though somewhat out of date, is still an excellent introduction to the subject. It contains valuable details on the topics summarized in Sections 2, 3 and 4 of this chapter.

7.1 An authoritative account of diffusional limitation in homogeneous reactions is treated by R. M. Noyes in *Progress in Chemical Kinetics*, Vol. I, G. Porter, ed. Pergamon Press, Inc., Long Island City, N.Y., 1961. The illustration of the cage effect by photolysis of azomethane is due to R. K. Lyon and D. H. Levy, *J. Am. Chem. Soc.*, **83**, 4290 (1961). Many aspects of these questions are covered in E. F. Caldin's *Fast Reactions in Solutions*, John Wiley and Sons, Inc., New York, 1964.

7.2 Further on the mathematical treatment of chain reactions and wall effects is found in Dainton's *Chain Reactions*, John Wiley and Sons, Inc., New York, 1956.

7.3 A now classical treatment of diffusional phenomena in porous catalysts is that of Ahlborn Wheeler, in *Catalysis*, Emmett, ed., Vol. 2, Reinhold Publishing Corp., New York, 1955.

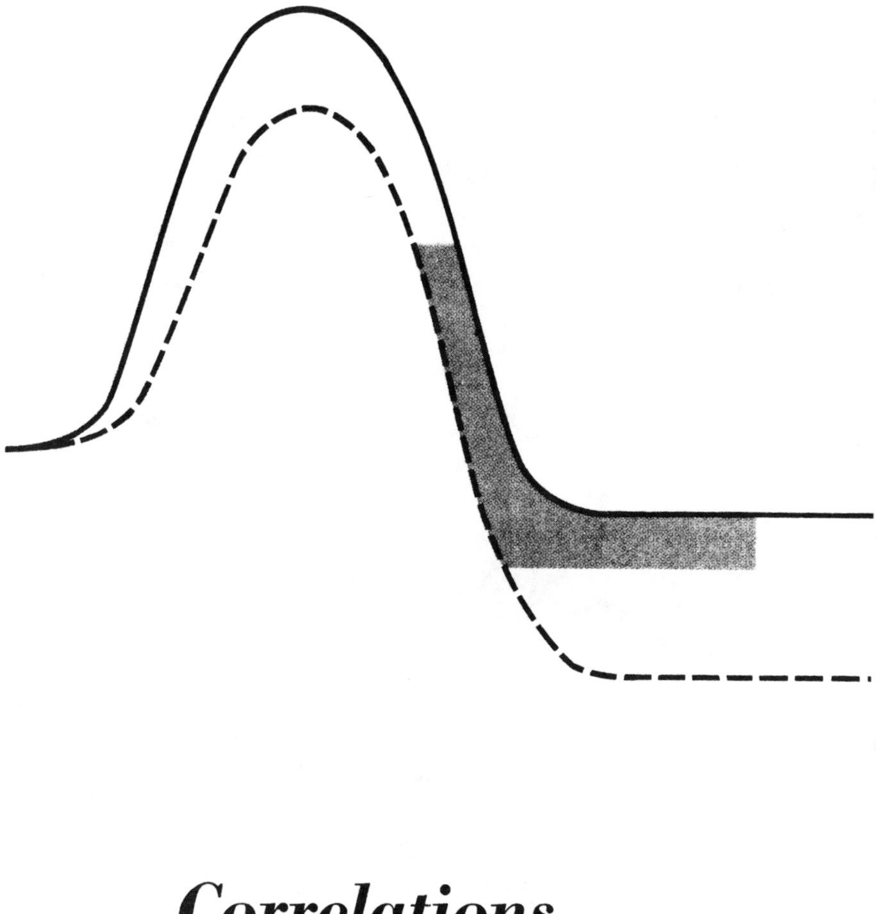

Correlations in Homogeneous Kinetics

8

It is frequently stated that kinetics and thermodynamics are two completely different subjects. This is an oversimplification which is misleading if it conveys the impression that thermodynamics is useless in the study of reaction rates. First of all, as was seen in Chapter 2, the theory of reaction rates is essentially an equilibrium theory. It permits us to evaluate pre-exponential factors of rate constants of elementary steps, at least in order of magnitude. As to the estimation of activation barriers, this can be done in certain cases, as will be seen shortly, by means of correlations between activation barriers and heats of reactions. Finally, in the treatment of reactions in concentrated acid solutions, it is frequently possible to relate rate constants to a purely thermodynamic quantity, the acidity function, which will also be introduced in this chapter.

Some simple empirical approaches will now be presented. Partial success in many situations should make it clear that the great stumbling block in the evaluation of reaction rates is not so much ignorance of rate constants of elementary steps but rather the dissection of the reaction into the appropriate sequence of its elementary steps. Once this dissection is made correctly, a fair estimate of the rate becomes possible at least under favorable circumstances. Such estimates are extremely valuable, not because they can ever be used with any degree of confidence in theoretical work but because agreement or disagreement with observed rates will confirm or refute the validity of the postulated sequence. A correct sequence is also very useful in applied work because it leads to the correct rate equation that is indispensable in the optimization of the reaction and of the reactor.

8.1 The Polanyi Relation

Thermodynamic correlations are frequently expected in homologous series of closely related compounds. The analog of a homologous series in chemical kinetics is a family of reactions. In such a family, the elementary steps will be characterized by potential energy surfaces which will exhibit roughly the same shape (Fig. 8.1.1). In particular, from the shape of the curves, it is not

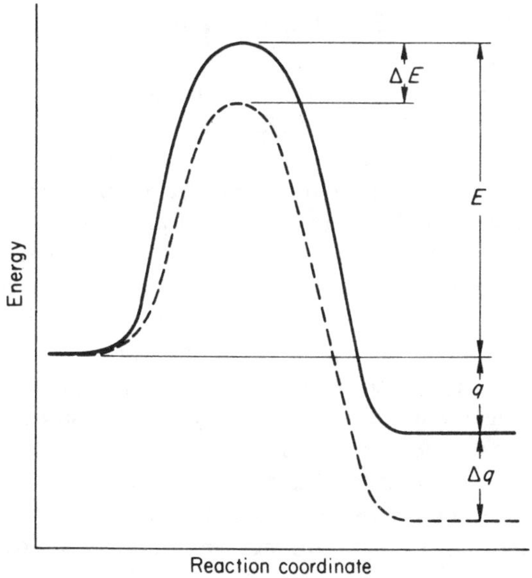

Fig. 8.1.1 The Polanyi relation.

unreasonable to find a proportionality between changes in activation barriers E and heats of reaction q:

$$\boxed{\Delta E = -\alpha \Delta q} \tag{8.1.1}$$

This relation, which is associated with the name of Polanyi (1935), is semi-empirical and cannot be proved satisfactorily. It says the following: If the heat of reaction q (considered as positive for an exothermic step) increases by an amount Δq from one member of the family to the next, then the activation barrier E will decrease by an amount ΔE equal to a fraction α of the change Δq, this fraction being between zero and unity.

Naturally, the parameter α may change from family to family. So will the parameter E_0 in the equivalent form of the *Polanyi relation* which also applies to exothermic elementary steps:

$$\boxed{E = E_0 - \alpha q} \tag{8.1.2}$$

Nevertheless, for a relatively large number of elementary steps involving reaction between a molecule and a free radical, the following values of the parameters:

$$\begin{aligned} \alpha &= 0.25 \\ E_0 &= 11.5 \text{ kcal/g-mole} \end{aligned} \tag{8.1.3}$$

provide, as shown by Semenov (1958), a useful guide and a means to estimate the activation energy if the heat of reaction is known.

But how is it possible to determine the heat of reaction for a step involving free radicals? In general, of course, heats of reaction are calculated from tabulated heats of formation of reactants and products. These tables rarely include the free radicals that appear as active centers in catalytic sequences.

The necessary information is more frequently given in tables of bond dissociation energies. A bond dissociation D(A—B) is defined as the energy required to rupture a bond linking the two atoms or molecular groups A and B in a molecule AB. It must be noted that for example the strength of a C—H bond in a molecule is not at all a constant but depends strongly on the groups to which the carbon atom is attached. Thus the bond dissociation energy $D(H{-}CH_3)$ in methane is 101 kcal/g-mole while $D(CH_3C_6H_4CH_2{-}H)$ in orthoxylene is only 74 kcal/g-mole. It takes 27 kcal less to break a C—H bond in the latter molecule than in methane. From bond dissociation energies, the heat of reaction is readily calculated. As an illustration, take the branching step in the $H_2{-}O_2$ system:

$$\mathbf{H + O_2 \rightarrow OH + O}$$

$$q = -D(O{-}O) + D(O{-}H) = -118 + 103 = -15 \text{ kcal/g-mole}$$

Thus, the step is endothermic: Its heat of reaction is -15 kcal/g-mole.

A useful empirical formula has been proposed by Van Tiggelen (1965) to estimate the bond dissociation energy $D(\text{A—B})$ where A and B may be atoms or radicals. Accordinging to this formula:

$$D(\text{A—B}) = \frac{\gamma_A D_A + \gamma_B D_B}{\gamma} \qquad (8.1.4)$$

where γ_A and γ_B are constants characteristic of radicals A and B, γ stands for either γ_A or γ_B, whichever has the smaller value, and D_A and D_B denote $D(\text{A—A})/2$ and $D(\text{B—B})/2$ respectively. In Table 8.1.1, values of γ_R and $\gamma_R D_R$ are collected for 35 different radicals R. Bond dissociation energies for a very large number of molecules are readily calculated by means of this table. For instance, let us calculate the dissociation energy of the carbon-carbon bond in $CH_3CO—C_2H_5$. Since $\gamma_{CH_3CO} < \gamma_{C_2H_5}$, and also $\gamma_{CH_3CO} = 1$, we have:

$$D(CH_3CO - C_2H_5) = \frac{\gamma_{CH_3CO} D_{CH_3CO} + \gamma_{C_2H_5} D_{C_2H_5}}{\gamma_{CH_3CO}}$$

$$= 31.5 + 46 = 77.5 \text{ kcal/g-mole}$$

The result is well within the experimental error of the measured value, namely 77.0 kcal/g-mole.

As an illustration of the use of Polanyi's relation combined with a means of estimating bond dissociation energies, a number of activation energies have been calculated by means of Eq. (8.1.2) and Table 8.1.1. The agreement

Table 8.1.1

VALUES OF γ_R AND $\gamma_R D_R$ (in kcal/g-mole) FOR VARIOUS RADICALS R

R	γ_R	$\gamma_R D_R$	R	γ_R	$\gamma_R D_R$
H	1.00	51.6	CH_2Cl	1.22	45.8
CHO	1.00	26.5	CH_2Br	1.23	43.4
CH_3CO	1.00	31.5	CBr_3	1.24	28.4
C_6H_5CO	1.00	23.0	$CHBr_2$	1.25	36.4
CN	1.05	60.0	$CHCl_2$	1.25	41.6
t-C_4H_9	1.05	31.5	CCl_3	1.29	37.4
$CH\equiv C$	1.06	69.4	CF_3	1.29	51.4
i-C_3H_7	1.10	36.5	$CH_2=CH-CH_2$	1.32	24.4
SH	1.15	39.1	Br	1.48	33.7
C_6H_5	1.18	60.0	NO	1.60	7.7
C_2H_5	1.18	46.0	NH_2	1.70	49.5
$CH_2=CH$	1.19	52.4	Cl	1.75	50.0
CH_3	1.20	49.8	NO_2	2.00	16.0
n-C_3H_7	1.20	43.3	OH	2.45	62.5
n-C_4H_9	1.20	44.4	CH_3O	2.95	54.5
$C_6H_5CH_2$	1.20	28.2	C_2H_5O	3.33	52.5
I	1.20	21.4	F	4.57	82.4

with experimental values is excellent, as shown by Table 8.1.2. It is essential to note that these considerations have only a limited applicability to certain families of reactions. But the success obtained in several instances is at least encouraging.

While it is still difficult to predict whether the Polanyi relation shall be valid for a particular reaction in a given family, it is perhaps easier to guess that the relation will not work in all cases where *steric factors* are likely to be important, in particular when steric hindrance is an essential contributor to the energy of the transition state.

For instance, *abstraction* reactions of the type:

$$\mathrm{Na + ClR \rightarrow NaCl + R}$$

that were studied extensively by Polanyi and his school provide a classical example of applicability of the Polanyi relation: The sodium atom can approach the chlorine atom attached to a carbon atom without being hindered

Table 8.1.2

Application of the Polanyi Relation to Reactions Between Molecules and Free Radicals

Reaction	Activation Energy Calculated from Table 8.1.1 and Eq. (8.1.2) With Values (8.1.3) (in kcal/g-mole)	Experimental Value of the Activation Energy* (in kcal/g-mole)
$\mathbf{H} + CH_4 \rightarrow H_2 + \mathbf{CH_3}$	11.0	13.0
$\mathbf{H} + C_2H_6 \rightarrow H_2 + \mathbf{C_2H_5}$	10.1	9.5
$\mathbf{H} + C_3H_8 \rightarrow H_2 + \boldsymbol{i}\text{-}\mathbf{C_3H_7}$	7.7	8.5
$\mathbf{H} + CH_3CHO \rightarrow H_2 + \mathbf{CH_3CO}$	6.5	6.0
$\mathbf{H} + CCl_4 \rightarrow HCl + \mathbf{CCl_3}$	3.0	3.5
$\mathbf{OH} + CH_4 \rightarrow H_2O + \mathbf{CH_3}$	8.3	8.5
$\mathbf{OH} + CH_3CHO \rightarrow H_2O + \mathbf{CH_3CO}$	3.8	4.0
$\mathbf{CH_3} + C_2H_6 \rightarrow CH_4 + \mathbf{C_2H_5}$	10.5	10.4
$\mathbf{CH_3} + CH_3CHO \rightarrow CH_4 + \mathbf{CH_3CO}$	7.0	6.8

*From *A Course of Chemical Kinetics* by N. M. Emanuel and D. G. Knorre, Moscow, 1962 (in Russian).

by the other three substituents attached to that carbon. On the contrary, in *substitution* reactions of the type:

$$I^- + RCl \rightarrow IR + Cl^-$$

occurring through the transition state pictured on the cover of this book (see also Fig. 1.1.1), the large entering group I^- attacks the carbon atom attached to the leaving group Cl^- on the opposite side from the latter and as the configuration of the carbon atom is inverted in the substitution, the other substituents attached to the carbon atom have to make room for the entering group. The size (steric hindrance) of the substituents becomes essential in determining changes in activation energies. As a result, the Polanyi relation is not expected to hold. The importance of steric factors in such substitution reactions was also anticipated by Polanyi as early as in 1931.

Problem 8.1.1

Write down the form of Polanyi's relation (8.1.2) as applicable to an endothermic step.

Problem 8.1.2

Estimate the activation energy of the branching step of the H_2—O_2 system.

Problem 8.1.3

Estimate the activation energy of the elementary step:

$$\mathbf{C_2H_5} + H_2 \rightarrow C_2H_6 + \mathbf{H}$$

Problem 8.1.4

Calculate the ratio:

$$(CH_2O)_{max}/(CH_4)_0$$

in a methane-air mixture at 700°K (5.4.5). The following value is supplied:

$$D(H\text{—}O_2H) = 91 \text{ kcal/g-mole.}$$

The result will verify an assumption made in deriving (5.4.5), namely that the maximum concentration of formaldehyde is reached at the early stages of the reaction (small conversions).

Problem 8.1.5

In order to verify a statement made on page 134 concerning the ratio of primary and secondary (degenerate branching) initiation, calculate this ratio at 700°K knowing that $D(\text{H—O}_2) = 45$ kcal/g-mole. Assume that the reacting mixture contains one molecule of CH_2O per 10^4 molecules of methane.

Problem 8.1.6

Using (6.3.17) estimate the maximum rate of oxidation of methane in an equimolar mixture of methane and oxygen at a total pressure of $\frac{1}{2}$ atm and 700°K. Assume that the chain length is equal to 100. This latter assumption cannot be justified rigorously, but is not worse than the educated guesses that must be made concerning probability factors. Clearly the end result may differ by several orders of magnitude from the experimental value.

Problem 8.1.7

A sequence was proposed in Problem 3.4.4 for the pyrolysis of acetaldehyde. Predict the apparent activation energy of the overall reaction.

Find reuuired bond dissociation energies in Table 8.1.1. It is further known that $D(\text{CH}_3\text{—CO}) = 12$ kcal/g-mole.

8.2 The Brönsted Relation

So far, the Polanyi relation has been applied only to catalytic sequences with free radicals as intermediates. But its scope is much wider. Many organic reactions in the liquid phase proceed through ionic intermediates and are catalyzed by acids and bases. Before the Polanyi relation is shown to correlate rate constants in acid-base catalysis, it will be put in an equivalent form by a series of self-explanatory steps. The temperature is assumed to be constant.

$$E = \text{const} - \alpha q \tag{8.1.2}$$

$$\frac{E}{RT} = \text{const} - \alpha \frac{q}{RT} \tag{8.2.1}$$

$$\exp\left(-\frac{E}{RT}\right) = \text{const} \times \exp\left(\frac{\alpha q}{RT}\right) \tag{8.2.2}$$

$$k = \text{const} \times K^{\alpha} \tag{8.2.3}$$

In order to proceed from (8.2.2) to (8.2.3), *it is necessary to assume that entropy factors, both thermodynamic and kinetic, do not change from one member of the family to the next*, so that they can be absorbed in a constant factor of proportionality. Then, Eq. (8.2.3) says that, for a family of reactions, rate constants k are proportional to equilibrium constants K elevated to some constant power α $(0 < \alpha < 1)$. Both α and the proportionality constant are characteristic of the family. Equation (8.2.3) is another form of Polanyi's relation.

The way it is applied to reactions catalyzed by acids or bases is best illustrated by an example, the acid-catalyzed dehydration of acetaldehyde hydrate:

$$\underbrace{CH_3-\overset{\displaystyle OH}{\underset{\displaystyle OH}{CH}}}_{S} \quad \rightarrow \quad \underbrace{CH_3-\overset{\displaystyle H}{CO}}_{P} + \underbrace{H_2O}_{W}$$

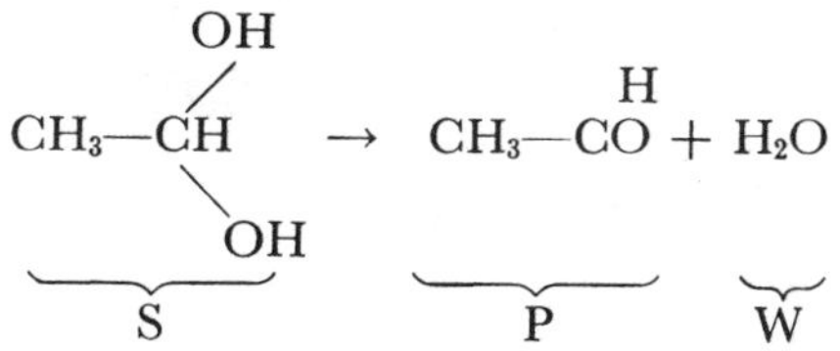

The catalytic sequence with an acid HA appears to be:

$$HA + S \;\leftrightarrow\; A^- + SH^+ \qquad (1)$$

$$SH^+ + A^- \;\rightarrow\; HA + S' \qquad (2)$$

$$S' \;\rightarrow\; P + W \qquad (3)$$

where the species S′ is

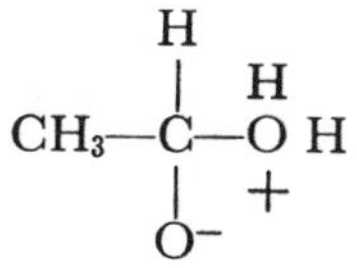

$$CH_3-\overset{\displaystyle H}{\underset{\displaystyle \underset{\displaystyle O^-}{|}}{\overset{\displaystyle |}{C}}}-\overset{\displaystyle H}{O}\; H \atop +$$

and **SH⁺** is the rate-determining active center.

Then the rate is equal to:

$$r = k_2(SH^+)(A^-)$$

The concentration of **SH⁺** is determined as usual from the condition of equilibrium for step (1):

$$(SH^+) = K_1 \frac{(S)(HA)}{(A^-)}$$

where K_1 is the equilibrium constant of step (1).

Therefore:

$$r = k_2 K_1(S)(HA) \qquad (8.2.4)$$

The rate is proportional to the concentration of the catalyst HA.

On the other hand, we may apply to step (2) the relation of Polanyi in the form (8.2.3)

$$k_2 = \text{const} \times K_2^{\alpha'} \tag{8.2.5}$$

where K' is the equilibrium constant of step (2) and α' is used instead of α for a reason that will be apparent shortly. This relation is expected to hold when various acids HA are used, step (2) with a given acid being a member of a family of reactions. Substitution of (8.2.5) into (8.2.4) gives:

$$r = \text{const} \times K_1 K_2^{\alpha'}(\text{S})(\text{HA}) \tag{8.2.6}$$

Consider now the dissociation constant of the acid HA in water corresponding to the equilibrium:

$$H_2O + HA \leftrightarrow H_3O^+ + A^- \qquad (4)$$

The equilibrium constant of (4) is

$$K_A = \frac{(H_3O^+)(A^-)}{(H_2O)(HA)} \tag{8.2.7}$$

The negative of the logarithm of the acidity constant K_A is called the pK of the acid:

$$pK_A = -\log K_A \tag{8.2.8}$$

Subtracting equilibria (1) and (4) side by side, we get:

$$S + H_3O^+ \leftrightarrow SH^+ + H_2O \qquad (5)$$

The equilibrium constant K_5 of (5) is simply equal to K_1/K_A. But K_5 is the acidity constant of S itself and does not depend on the nature of the acid used as catalyst. Therefore,

$$K_1 = \text{const} \times K_A \tag{8.2.9}$$

Similarly, adding up (2) and (4) side by side, we get:

$$SH^+ + H_2O \leftrightarrow S' + H_3O^+ \qquad (6)$$

with an equilibrium constant $K_6 = K_2 K_A$, a constant which does not depend on the nature of the acid used as catalyst.

Thus:

$$K_2 = \frac{\text{const}}{K_A} \tag{8.2.10}$$

Finally, substitution of (8.2.9) and (8.2.10) into (8.2.6) gives:

$$r = \text{const} \times K_A^{1-\alpha'}(S)(HA) \tag{8.2.11}$$

It is seen that the rate constant k_A of the acid-catalyzed reaction:

$$r = k_A(S)(HA)$$

is of the form:

$$\boxed{k_A = \text{const} \times K_A^{\alpha}} \tag{8.2.12}$$

where $\alpha = 1 - \alpha'$ is a constant: $0 < \alpha < 1$.

Equation (8.2.12) is known as the Brönsted relation (Fig. 8.2.1). It was

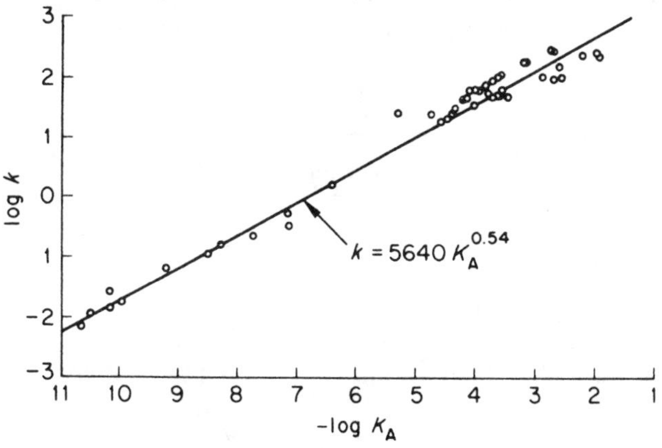

Fig. 8.2.1 Illustration of the Brönsted relation. Rate constants for the acid-catalyzed dehydration of acetaldehyde hydrate vs dissociation constants for 44 different catalytic carboxylic acids and phenols. [Bell & Higginson, *Proc. Roy. Soc.*, **A197**, 150 (1949)]

first proposed by Brönsted and Pedersen in 1924. It says that the rate constant of a reaction catalyzed by an acid is proportional to some power of the dissociation constant of the acid used as catalyst.

Similarly for a reaction catalyzed by a base:

$$\boxed{k_B = \text{const} \times K_B{}^{\beta}} \qquad (8.2.13)$$

where k_B is the rate constant and K_B is the dissociation constant of the base.

The important point about the illustration is that the Brönsted relation is seen to be a straightforward consequence of the Polanyi relation. These empirical relations are very general, and do not depend on the details of the catalytic sequence or the nature of its rate-determining step. Their usefulness lies in the possibility of predicting a rate from a purely thermodynamic quantity, K_A, provided that the proportionality constant and α have been determined by means of experiments with a series of acids (or bases).

Problem 8.2.1

G. N. Lewis and G. T. Seaborg have determined rate constants k (in liter mole^{-1} sec^{-1}) for the neutralization of the anion of p - p' - p'' - trinitrotriphenylmethane by a number of weak acids in alcohol solution at $-60°C$. The pK values of these acids are known and may be used to correlate the rate constants in Brönsted fashion. The data are:

Acid	pK	k
monochloroacetic	2.80	20
furoic	3.15	10
α-naphthoic	3.70	5.2
benzoic	4.15	5.0
acetic	4.75	2.9
p-nitrophenol	7.10	0.9
hydrocyanic	9.15	0.05
β-naphthol	9.60	0.03
phenol	9.85	0.01

Predict the value of the rate constant for 2,4-dichlorophenol with a pK value equal to 7.55.

8.3 The Hammett Relation: Linear Free-energy Relationships

Consider a family Φ of reactions: the esterification of ethyl alcohol by benzoic acid and its derivatives obtained by introducing substituents in various positions of the benzene ring, e.g.:

$$\left.\begin{array}{rcl} C_6H_5COOH + C_2H_5OH & \rightarrow & C_6H_5COOC_2H_5 + H_2O \\ m\text{-}NO_2C_6H_4COOH + C_2H_5OH & \rightarrow & m\text{-}NO_2C_6H_4COOC_2H_5 + H_2O \\ p\text{-}ClC_6H_4COOH + C_2H_5OH & \rightarrow & p\text{-}ClC_6H_4COOC_2H_5 + H_2O \end{array}\right\} \Phi$$

For this family, we expect a Polanyi relation in the form (8.2.3) *if the thermodynamic and kinetic entropies do not vary* from one member of the family to the next. This condition is found to be satisfied unless substituents are in ortho positions, in which case steric effects introduce changes in the entropy factors. Let us rewrite (8.2.3) as:

$$k = \text{const } K^{\rho} \tag{8.3.1}$$

where the exponent $0 < \rho < 1$ is called ρ and not α by tradition.

In particular for the reaction with benzoic acid (no substituent), taken as reference reaction (subscript 0):

$$k_0 = \text{const } K_0^{\rho} \tag{8.3.2}$$

Therefore, from (8.3.1) and (8.3.2), we obtain:

$$\log k - \log k_0 = \rho\,[\log K - \log K_0] \tag{8.3.3}$$

Consider now a second family Φ^* of reactions related to the one just considered. Many are possible. The simplest one will be the family obtained by replacing ethyl alcohol by water, i.e., a family of the ionization reactions of benzoic acids in water:

$$\left.\begin{array}{rcl} C_6H_5COOH + H_2O & \rightarrow & C_6H_5COO^- + H_3O^+ \\ m\text{-}NO_2C_6H_4COOH + H_2O & \rightarrow & m\text{-}NO_2C_6H_4COO^- + H_3O^+ \\ p\text{-}ClC_6H_4COOH + H_2O & \rightarrow & p\text{-}ClC_6H_4COO^- + H_3O^+ \end{array}\right\} \Phi^*$$

Let us take Φ^* as a reference family of reactions. Following Hammett, we now postulate a *linear relationship between standard free energies:*

$$\underbrace{\Delta G - \Delta G_0}_{\text{Family } \Phi} = \rho_K\, (\underbrace{\Delta G^* - \Delta G_0^*}_{\text{Family } \Phi^*}) \tag{8.3.4}$$

This relation says that the change in free energy ΔG for a reaction in a given family (referred to the change in free energy for the reaction involving the unsubstituted compound ΔG_0) is proportional to the change in free energy ΔG^* for the corresponding reaction in the reference family (referred to the change in free energy for the reaction of the reference family involving the unsubstituted compound ΔG_0^*).

In general, linear free-energy relationships of the type (8.3.4) cannot be justified. When used in connection with Polanyi's relation, they lead to a very useful system of empirical correlations. Indeed, (8.3.4) can be rewritten in terms of equilibrium constants K with the same subscripts and superscripts as used for ΔG:

$$\log K - \log K_0 = \rho_K (\log K^* - \log K_0^*) \tag{8.3.5}$$

Now, $\log K^* - \log K_0^*$ is a difference that depends only on the nature and position of the substituent since the corresponding reactions are those of the reference family. Thus:

$$\log K^* - \log K_0^* = \sigma \tag{8.3.6}$$

where σ is a constant characterizing the nature and position of the substituent.

Substitution of (8.3.6) into (8.3.5) gives:

$$\log \frac{K}{K_0} = \rho_K \sigma \tag{8.3.7}$$

Combining (8.3.3) and (8.3.7), we get

$$\log \frac{k}{k_0} = \rho \rho_K \sigma \tag{8.3.8}$$

Combining the constants: $\rho \rho_K = \rho_k$, we obtain finally the *Hammett relation:*

$$\boxed{\log \frac{k}{k_0} = \rho_k \sigma} \tag{8.3.9}$$

The constant $\rho \rho_K = \rho_k$ depends only on the nature of the reaction.

The Hammett relation permits us to predict the rate constant of a given reaction involving a molecule with a certain substituent provided that the constants ρ_k and σ characteristic of that reaction and that substituent are known.

Hammett's relations have been refined and extended to many types of reactions. These semi-empirical correlations and their many variations have, of course, great theoretical importance and their interpretation constitutes a lively field of organic chemistry. From the viewpoint of applied chemical kinetics, they may be of use in practical situations, since they help to predict an unknown rate constant. However, more important is the philosophy behind these correlations; it provides a very general technique for bridging the gap between thermodynamic and kinetic properties. All it involves is the Polanyi relation plus a linear free-energy relationship.

Problem 8.3.1

Show that the constant ρ in Hammett's relation (8.3.9) is expected to be inversely proportional to absolute temperature.

Problem 8.3.2

Consider a family of reactions:

$$\mathrm{R} + \mathrm{R}_i\mathrm{X} \quad \rightarrow \quad \mathrm{RX} + \mathrm{R}_i$$

Within the family, the free radical R changes while $\mathrm{R}_i\mathrm{X}$ remains the same. Another family of reactions might be:

$$\mathrm{R} + \mathrm{R}_i'\mathrm{X} \quad \rightarrow \quad \mathrm{RX} + \mathrm{R}_i'$$

which differs from the first one by the fact that R_i has been replaced by R_i'.

In each family, the reference reaction characterized by a rate constant k_0 will be a reaction with a radical R_0, e.g.:

$$R_0 + \mathrm{R}_i\mathrm{X} \quad \rightarrow \quad \mathrm{R}_0\mathrm{X} + \mathrm{R}_i$$

Derive a Hammett relation for these families of reaction, making use of (8.1.3) and assuming that γ_{X} is smaller than γ_{R} and γ_{R_i}. What is the significance of ρ_k and σ in the relation found? Assume that entropies of reaction do not change within a family and from one family to the next.

8.4 *The Compensation Effect*

In a family of reactions, the rate constants can generally be put in the Arrhenius form:

$$k = A \exp\left(-\frac{E}{RT}\right) \tag{1.4.2}$$

If the reactions are elementary steps, we can write for the case of thermodynamically ideal systems [see (2.4.26)]

$$k_0 = \frac{kT}{h} \exp\left(\frac{\Delta S^{0\ddagger}}{R}\right) \exp\left(-\frac{\Delta H^{0\ddagger}}{RT}\right) \tag{8.4.1}$$

In the correlations considered thus far, namely those associated with the names of Hammett and Brönsted, differences in rate constants k_0 were attributed to differences in activation energies E (or $\Delta H^{0\ddagger}$) while pre-exponential factors A (or $\Delta S^{0\ddagger}$) were assumed to be constant from one member of the family to the next. The opposite case (constant E, varying A) is also known, when, for instance, a reaction is run in different solvents.

The situation where both A and E change simultaneously is, however, quite frequent. In liquid phase reactions in particular, it happens that A and E change in such a way that the corresponding change in the rate constant is much less than would be the case if only E or A changed alone. This phenomenon is called the *compensation effect* and it has already been alluded to (see p. 46). In its most striking form, the compensation effect for the reactions of a given family can be expressed by a linear relation:

$$\ln A = (\text{const})_1 + (\text{const})_2 E \tag{8.4.2}$$

If the reactions are elementary steps, it is permissible to rewrite (8.4.1) as follows:

$$\left(\frac{\Delta S^{0\ddagger}}{R}\right) = (\text{const})_1 + (\text{const})_2\, \Delta H^{0\ddagger} \tag{8.4.3}$$

The constants in (8.4.2) and (8.4.3) are not the same but the same notation is used for the sake of simplicity. Since terms on the left-hand side of Eqs. (8.4.2) and (8.4.3) are dimensionless, the second parameter on the right-hand side of these equations must have the dimensions of a reciprocal energy per mole. It is therefore tempting to write:

$$(\text{const})_2 = \frac{1}{R\theta} \tag{8.4.4}$$

where θ is a new parameter with the meaning of a temperature. Substitution of (8.4.2) and (8.4.3) into (1.4.2) and (8.4.1), respectively, gives:

$$k = A' \exp\left[-\frac{E}{R}\left(\frac{1}{T} - \frac{1}{\theta}\right)\right] \tag{8.4.5}$$

$$k_0 = A'' \exp\left[-\frac{\Delta H^{0\ddagger}}{R}\left(\frac{1}{T} - \frac{1}{\theta}\right)\right] \tag{8.4.6}$$

In these expressions, use has been made of (8.4.4); A' and A'' are constants for the reactions or elementary steps of a given family. Therefore the meaning of θ is that of an *isokinetic temperature*, i.e., a temperature at which rate constants for all reactions or elementary steps of a family have the same value. Below that temperature, the reactions with a smaller activation energy are faster. Above the isokinetic temperature, the faster reactions are the ones with the larger values of the activation energy. These relations are illustrated in the Arrhenius diagram of Fig. 8.4.1.

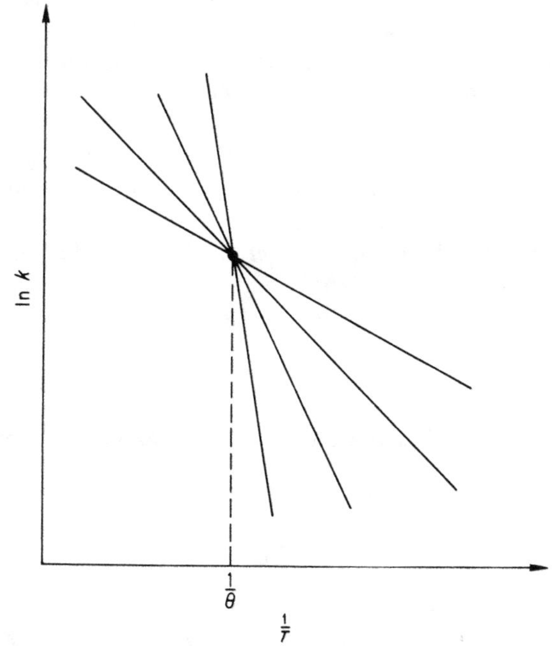

Fig. 8.4.1 The compensation effect.

If the isokinetic temperature θ is right in the middle of the restricted range of temperatures inside which measurements are being made, the compensation effect, in the form of (8.4.2) or (8.4.3), may well be simply a reflection of experimental errors.

On the other hand, if θ is far above or below the range of temperatures covered by the measurements, and if the variation in activation energies is large, the linear relations (8.4.2) or (8.4.3) have a definite meaning. From a practical standpoint, they express a useful correlation. Ultimately, they may well receive a theoretical explanation although at the moment they must be considered as empirical like the Brönsted and Hammett relations. Finally, the occurrence of compensation effects makes it imperative to measure rate constants not merely at one temperature but over as wide a temperature interval as possible. Otherwise, the loss of information can lead to serious errors concerning order of reactivities, as can be seen from Fig. 8.4.1, which shows the inversion of reactivities at the isokinetic temperature.

8.5 The Acidity Function in Acid-Base Catalysis

As pointed out in Chapter 2, although theory can account in principle for kinetic behavior in systems that are nonideal from a thermodynamic standpoint, corrections for nonideality are generally ignored in chemical kinetics at the present time. A notorious exception concerns acid-base catalysis in concentrated acid solutions.

Indeed, following Hammett, the acidity of such solutions, as measured by its catalytic effects, can be described remarkably well by an acidity function H_0 which takes the place of pH in concentrated solutions.

To define this acidity function, consider any base B and its conjugate acid BH^+. Let us first introduce a quantity h_0 defined by the relation:

$$h_0 = K_{BH^+} \frac{(BH^+)}{(B)} \tag{8.5.1}$$

where K_{BH^+} is the thermodynamic dissociation constant of BH^+ corresponding to the equilibrium $BH^+ \rightleftarrows B + H^+$. This constant generally must be written in terms of activities:

$$K_{BH^+} = \frac{a_B a_{H^+}}{a_{BH^+}} = \frac{1}{K_B} \tag{8.5.2}$$

except in sufficiently dilute solutions that behave ideally. Then:

$$K_{BH^+} = \frac{(B)(H^+)}{(BH^+)}$$

Remembering (8.2.8), we then introduce *the acidity function* H_0 by the definition:

$$H_0 = -\log h_0 = pK_{BH^+} - \log \frac{(BH^+)}{(B)} \tag{8.5.3}$$

It is readily seen that in an ideal solution, H_0 reduces to:

$$H_0 = -\log (H^+) = pH \tag{8.5.4}$$

The reason why H_0 is a useful quantity is that, as can be shown experimentally, it has a value for a given acid solution that is approximately independent of the base B used.

The usefulness of the acidity function in kinetic problems is best seen by means of an example, the acid-catalyzed lactonization of γ-hydroxybutyric acid (B) into γ-butyrolactone (L):

$$
\begin{array}{c}
\text{CH}_2\text{—COOH} \\
\diagup \\
\text{CH}_2 \\
\diagdown \\
\text{CH}_2\text{—OH}
\end{array}
\quad \rightleftarrows \quad
\begin{array}{c}
\text{CH}_2\text{—CO} \\
\diagup \qquad\; | \\
\text{CH}_2 \qquad\; \\
\diagdown \qquad\; | \\
\text{CH}_2\text{—O}
\end{array}
\quad + \text{H}_2\text{O}
$$

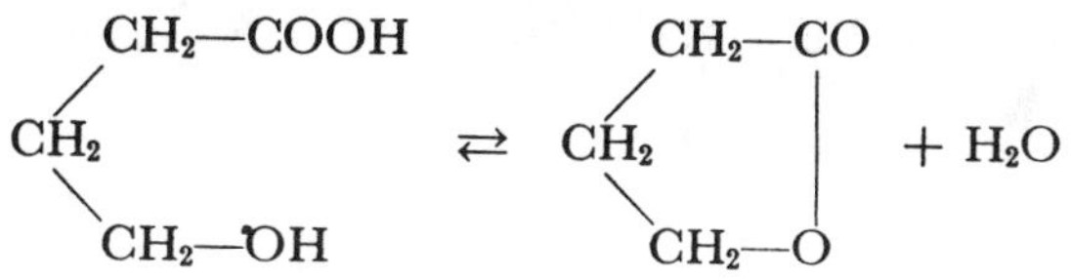

and the reverse reaction. The catalytic sequence can be written as follows:

$$\text{B} + \text{H}^+ \;\leftrightarrow\; \text{BH}^+ \tag{1}$$

$$\text{BH}^+ \;\rightleftarrows\; \text{LH}^+ + \text{H}_2\text{O} \tag{2}$$

$$\text{LH}^+ \;\leftrightarrow\; \text{L} + \text{H}^+ \tag{3}$$

where step (2) is rate-determining. Steps (1) and (3) are in quasi-equilibrium.

We shall consider in turn the rates of the forward and reverse reactions. For the forward reaction:

$$\vec{r} = \vec{r}_2 = \vec{k}_2(\text{BH}^+)\,\frac{\gamma_{\text{BH}^+}}{\gamma^{\ddagger}} \tag{8.5.5}$$

where, as shown in Chapter 2, γ_{BH^+} is the activity coefficient of BH^+ and $\gamma^{\ddagger}$ is the activity coefficient of the transition state of step (2). Since the latter differs very little from the reacting molecule BH^+, it is reasonable to assume that $\gamma_{\text{BH}^+} = \gamma^{\ddagger}$. Then (8.5.5) becomes:

$$\vec{r}_2 = \vec{k}_2(\text{BH}^+) \tag{8.5.6}$$

The concentration of the active center BH^+ is determined as usual from the appropriate fast equilibrium (1) which can be expressed in terms of h_0 from (8.5.2) and (8.5.3):

$$(\text{BH}^+) = K_{\text{B}} h_0 (\text{B}) \tag{8.5.7}$$

Substitution of (8.5.7) into (8.5.6) gives:

$$\vec{r} = \vec{r}_2 = \vec{k}_2 K_{\text{B}} h_0 (\text{B}) = \vec{k}(\text{B}) \tag{8.5.8}$$

The rate constant for the forward reaction $\vec{k} = \vec{k}_2 K_B h_0$ *is proportional to* h_0 *and therefore* $-\log \vec{k}$ *must change linearly with the acidity function* H_0 *as is verified experimentally.*

The behavior of the reverse reaction is different:

$$\overleftarrow{r} = \overleftarrow{r}_2 = \overleftarrow{k}_2(LH^+)(H_2O)\frac{\gamma_{LH} + \gamma_{H_2O}}{\gamma^\ddagger} = \overleftarrow{k}_2\, a_{LH} + a_{H_2O}\frac{1}{\gamma^\ddagger} \qquad (8.5.9)$$

But, by definition:

$$K_L = \frac{a_{LH^+}}{a_L a_{H^+}} \quad \text{and} \quad K_w = \frac{a_{H_2O}a_{H^+}}{a_{H_3O}} \qquad (8.5.10)$$

Thus

$$\overleftarrow{r} = \overleftarrow{k}_2 K_L K_w a_L a_{H_3O^+}\frac{1}{\gamma^\ddagger}$$

or

$$\overleftarrow{r} = \overleftarrow{k}_2 K_L K_w \frac{\gamma_L(\gamma_{H_3O^+})}{\gamma^\ddagger}(L)(H_3O^+)$$

It is reasonable to assume that the ratio of activity coefficients $\gamma_L\gamma_{H_3O^+}/\gamma^\ddagger$ does not change with the composition of the solution since the species in the numerator have the same charge and contain the same elements as the transition state. If this is the case:

$$\overleftarrow{r} = \overleftarrow{k}(L)$$

and we see that the rate constant for a given acid solution $k = k'(H_3O^+)$ is proportional not to h_0 but simply to the concentration of the ion (H_3O^+). This is also verified experimentally.

In conclusion, the rate constant of acid-base catalyzed reactions may or may not be proportional to h_0. This depends on the details of the catalytic sequence. But when such a proportionality exists, the tabulated values of acidity functions for a large variety of concentrated acid solutions constitute a very valuable tool for kinetic correlations.

Problem 8.5.1

In a new process for the commercial production of isoprene, the first reaction involves the acid-catalyzed addition of an olefin R—CH═CH$_2$

to formaldehyde in concentrated solutions of sulfuric acid. The catalytic sequence is believed to be:

$$HCHO + H^+ \leftrightarrow {}^+CH_2OH \tag{1}$$

$$R{-}CH{=}CH_2 + {}^+CH_2OH \rightarrow R{-}{}^+CH{-}CH_2{-}CH_2OH \tag{2}$$

$$R{-}{}^+CH{-}CH_2{-}CH_2OH \rightarrow \text{Products} \tag{3}$$

Step (2) is rate-determining. Is the rate constant expected to be proportional to h_0? Justify your positive or negative answer.

BIBLIOGRAPHY

8.1 A recent review on "Bond Energies" has appeared under the name of S. W. Benson, in *J. Chem. Ed.*, **42**, 502 (1965). A critical compilation of bond dissociation energies is found in Cottrell's *The Strength of Chemical Bonds*, Butterworths, London, 1958. Application of Polanyi's relation to free radical reactions has been proposed by Semenov in *Some Problems in Chemical Kinetics and Reactivity*, translated by M. Boudart, Vol. I, Princeton University Press, Princeton, N.J., 1958. The empirical formula (8.1.4) is proposed by A. Van Tiggelen, J. Peeters and R. Burke, *Chem. Eng. Sci.*, **20**, 529 (1965). The results of an extensive survey of steric effects in substitution reactions have been summarized by C. K. Ingold, *Quart. Rev.*, **11**, 1 (1957).

8.2 The Brönsted relations, as presented here, constitute a brief introduction to a vast and well-organized chapter of chemical kinetics: acid-base catalysis. A recent survey of the field is *The Proton in Chemistry*, by R. P. Bell, Methuen & Co. Ltd., London, 1959.

8.3 The Hammett relations, as presented here, constitute a brief introduction to a vast, well-organized and rapidly-growing chapter of physical organic chemistry, with which are associated the names of Taft, Brown, Winstein, and others. See, for instance, Jack Hine, *Physical Organic Chemistry*, McGraw-Hill Book Company, New York, 1963. The constants σ and ρ_k of Hammett's relation have been tabulated. They can be found, for example, in Hammett's book *Physical Organic Chemistry*, pp. 188-91, McGraw-Hill Book Company, New York, 1940. Values of σ and ρ_k can be interpreted from the viewpoints of theories of chemical reactivity.

8.4 Although the compensation effect has been reported by many investigators in many areas of chemical kinetics over the past thirty years, its importance in homogeneous kinetics has been brought to the fore by Leffler [*J. Org. Chem.*, **20**, 1202 (1955)], who listed 81 families of reactions exhibiting an isokinetic temperature.

8.5 Values of H_0 functions for different acids in various solvents have been collected by F. A. Long and M. Paul, *Chem. Revs.*, **57**, 1 (1957). The example treated and illustrated in Fig. 8.4.1 is also borrowed from the work of Long, *et al.*

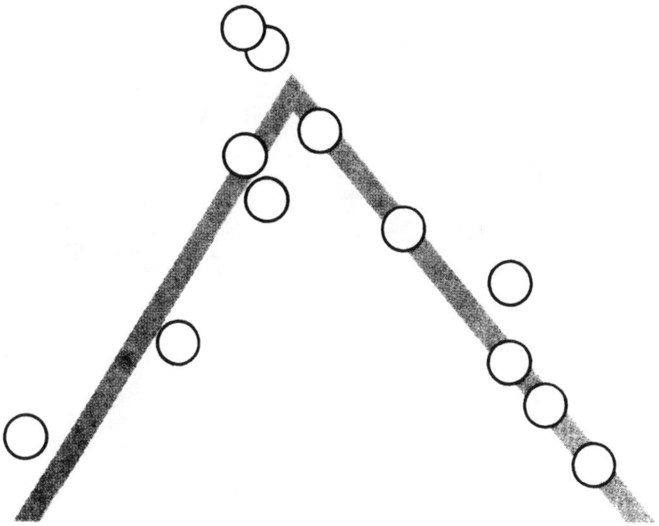

Correlations in Heterogeneous Catalysis

9

In the treatment of catalytic reactions presented in Chapters 3, 4 and 5, no distinction has been made between reactions taking place homogeneously and reactions catalyzed at solid surfaces. The latter now deserve special attention, not only because of their practical importance in the laboratory and in industry, but also because of the particular features of solid catalysts which tend to relegate the study of their catalytic action to a voluminous but separate topic of chemical kinetics.

The reason for this special treatment of *contact catalysis*, i.e., catalysis at solid surfaces, lies in the fact that although much is known about surface chemistry, little is known today about the active centers at the surface of solids. While considerable knowledge has been gained about the structure, thermodynamics and kinetics of many free radicals, not even the chemical identity of adsorption complexes, i.e., the active centers at solid surfaces, is known with certainty in any particular instance. The obstacles in the way of detailed understanding are further aggravated by the difficulty of reproducing identical batches of solid catalysts, even under very strict rules of preparation in a given laboratory. Traces of impurities that accumulate at the surface, changes in structure and texture of the solid prior to its use and during the course of reaction, make experimentation in this field particularly arduous. The rate functions obtained from kinetic data are difficult to interpret, since the rate and equilibrium constants of the elementary steps in the catalytic sequence are unknown and the elementary steps themselves have to be assumed without proof.

Nevertheless, the general rules that have been emphasized throughout this book certainly apply to contact catalysis. Our first concern will be to see how rate functions may be modified by the frequent if not universal occurrence on a given surface of active centers that differ in their thermodynamic and kinetic properties. This is the problem of *nonuniformity* of solid surfaces. Next, several *empirical correlations* that have emerged from kinetic studies will be considered briefly as an introduction to this special field of contact catalysis.

9.1 Catalysis on a Nonuniform Surface

Adsorption measurements generally reveal that the heat of formation of adsorption complexes varies significantly with surface coverage. Such a variation may serve as a good operational definition of *active sites*, the existence of which was first clearly formulated by Taylor in 1925. The consequences of the fact that not all sites at the surface of a solid (or of an enzyme molecule) may be equally active are many-sided. Thus, in certain situations, only very few selected sites may participate in the catalytic sequence; this is an extreme case of "active sites." In other situations, although most, if not all, of the sites take part in the reaction, they do so at different levels of activity.

In order to show the nature of the kinetic problem in a situation of this sort, we will consider the case where the nonuniformity of the active centers is approximated, at least in the domain of interest, by a linear variation with surface coverage of heats of formation of adsorption complexes. Consider a two-step catalytic sequence of the type treated on page 70:

$$\mathbf{S}_1 + A_1 \rightleftarrows B_1 + \mathbf{S}_2 \qquad (1)$$

$$\mathbf{S}_2 + A_2 \rightleftarrows B_2 + \mathbf{S}_1 \qquad (2)$$

where $\mathbf{S}_1$ and $\mathbf{S}_2$ denote two kinds of active centers and

$$(\mathbf{S}_1 + (\mathbf{S}_2) = (\mathbf{L})$$

as in Eq. (3.3.7).

Also, as usual, we write $(\mathbf{L}) = 1$ and therefore

$$(\mathbf{S}_1) + (\mathbf{S}_2) = 1 \tag{9.1.1}$$

Then, Eq. (3.3.9) gives directly the differential rate dr occurring on an infinitesimal fraction of the surface sites, ds, characterized by uniform properties of the active centers $\mathbf{S}_1$ and $\mathbf{S}_2$:

$$dr = \frac{k_1k_2(A_1)(A_2) - k_{-1}k_{-2}(B_1)(B_2)}{k_1(A_1) + k_2(A_2) + k_{-1}(B_1) + k_{-2}(B_2)}\, ds \tag{9.1.2}$$

In order to integrate (9.1.2), it is necessary to find how k_1, k_{-1}, k_2 and k_{-2} depend on the variable s.

For this purpose, let us assume that step (1) is exothermic from left to right. This is not a necessity and the whole reasoning may be repeated for the other alternative. The essential assumption is that the heat of reaction q_1 (counted as positive) depends linearly on s:

$$q_1 = q_1^0 - Cs \tag{9.1.3}$$

It is natural to assume that, as s changes, the Polanyi relation (8.1.2) applies to the activation barrier E_1 of step (1), from left to right. Hence:

$$E_1 = E_0 - \alpha q_1 \tag{9.1.4}$$

Since, furthermore, the activation energy for step (1) from right to left, E_{-1}, must be equal to $E_1 + q_1$, we have:

$$E_{-1} = E_0 + (1 - \alpha)q_1 \tag{9.1.5}$$

Using Arrhenius expressions (compare with Eq. 8.2.2) we then get from (9.1.3), (9.1.4) and (9.1.5)

$$k_1 = k_1^0 e^{-\alpha f s} \tag{9.1.6}$$

$$k_{-1} = k_{-1}^0 e^{(1-\alpha) f s} \tag{9.1.7}$$

where

$$f = \frac{C}{RT} \tag{9.1.8}$$

The constants k_1^0 and k_{-1}^0 are simply the values of the rate constants k_1 and k_{-1} for $s = 0$.

Similar expressions are obtained readily for k_2 and k_{-2}. Clearly, the heat q of the overall reaction

$$A_1 + A_2 \rightleftarrows B_1 + B_2$$

must not depend on s. As it was assumed that step (1), an adsorption step, is exothermic, it is reasonable to assume that step (2), a desorption step, is endothermic. Therefore:

$$q = q_1 - q_2$$

From (9.1.3), it follows that

$$q_2 = q_2^0 - Cs \tag{9.1.9}$$

But Polanyi's relation applies to exothermic steps, thus to step (2) from right to left, with an exothermic (positive) heat of reaction $-q_2$:

$$-q_2 = -q_2^0 + Cs \tag{9.1.10}$$

Polanyi's relation for step (2) is then:

$$E_{-2} = E_0 - \alpha(-q) \tag{9.1.11}$$

Consequently, we get for k_{-2} and k_2 relations similar to those for k_1 and k_{-1} respectively.

$$k_{-2} = k_{-2}^0 e^{-\alpha f s} \tag{9.1.12}$$

$$k_2 = k_2^0 e^{(1-\alpha) f s} \tag{9.1.13}$$

For integration of (9.1.2) it is convenient to use a variable u defined as:

$$u = \frac{k_{-1}(B_1) + k_2(A_2)}{k_1(A_1) + k_{-2}(B_2)} \tag{9.1.14}$$

With a value $u = u_0$ corresponding to $s = 0$, it is easily verified by means of (3.3.4) that

$$u = \frac{(S_1)}{(\mathring{S}_2)} \tag{9.1.15}$$

Substitution of (9.1.6), (9.1.7), (9.1.12) and (9.1.13) into (9.1.14) yields:

$$s = \frac{1}{f} \ln \frac{u}{u_0} \tag{9.1.16}$$

Differentiation of (9.1.16) gives:

$$ds = \frac{1}{f}\frac{du}{u} \tag{9.1.17}$$

Dividing both numerator and denominator of (9.1.2) by $k_1(A_1) + k_{-2}(B_2)$ and using (9.1.14), we get:

$$dr = \frac{k_1k_2(A_1)(A_2) - k_{-1}k_{-2}(B_1)(B_2)}{k_1(A_1) + k_{-2}(B_2)} \frac{ds}{1+u} \tag{9.1.18}$$

Substitution of (9.1.17) and use of the expressions for k_1, k_{-1}, k_2, k_{-2} transform (9.1.18) into:

$$dr = \frac{1}{f} \frac{k_1{}^0k_2{}^0(A_1)(A_2) - k_{-1}{}^0k_{-2}{}^0(B_1)(B_2)}{[k_1{}^0(A_1) + k_{-2}{}^0(B_2)]^\alpha [k_{-1}{}^0(B_1) + k_2{}^0(A_2)]^{1-\alpha}} \frac{u^{-\alpha}}{1+u} du \tag{9.1.19}$$

To integrate (9.1.19), we recall the definition (9.1.15) of u. The limits of integration $s = 0$ and $s = 1$ are replaced by $u = u_0$ and $u = u_0e^f$. At the lower limit, most of the sites will be occupied and therefore will be in the $\mathbf{S}_2$ form. Hence u_0 is a small quantity. On the other hand, because f is a large quantity (9.1.8) on a surface exhibiting marked nonuniformity, u_0e^f will be a large quantity or, in other words, at the other end of the spectrum, most sites are in the $\mathbf{S}_1$ form. Thus, within a good approximation, the limits of integration can be taken as $u = 0$ and $u = \infty$.

If we further note that:

$$\int_0^\infty \frac{u^{-\alpha}}{1+\mu} du = \frac{\pi}{\sin \alpha\pi} \tag{9.1.20}$$

integration of (9.1.19) gives finally:

$$\boxed{r = \frac{1}{f} \frac{\pi}{\sin \alpha\pi} \frac{k_1{}^0k_2{}^0(A_1)(A_2) - k_{-1}k_{-2}(B_1)(B_2)}{[k_1{}^0(A_1) + k_{-2}{}^0(B_2)]^\alpha [k_2{}^0(A_2) + k_{-1}{}^0(B_1)]^{1-\alpha}}} \tag{9.1.21}$$

The rate function (9.1.21) is not very different in form from the conventional one derived on the assumption of a uniform surface:

$$r = \frac{k_1k_2(A_1)(A_2) - k_{-1}k_{-2}(B_1)(B_2)}{k_1(A_1) + k_{-2}(B_2) + k_2(A_2) + k_{-1}(A_1)} \tag{9.1.22}$$

Both numerators in (9.1.21) and (9.1.22) are of the form of the law of mass action. The denominators are similar and both rate functions can be approximated by an expression of the type (4.2.8):

$$r = \vec{k}(A_1)^{\vec{\alpha}_1}(A_2)^{\vec{\alpha}_2}(B_1)^{\vec{\alpha}_3}(B_2)^{\vec{\alpha}_4} - \overleftarrow{k}(A_1)^{\overleftarrow{\alpha}_1}(A_2)^{\overleftarrow{\alpha}_2}(B_1)^{\overleftarrow{\alpha}_3}(B_2)^{\overleftarrow{\alpha}_4} \qquad (9.1.23)$$

where, as discussed on page 14, the exponents are now empirical but may stay constant over a restricted range of process variables.

Actually, our purpose in deriving (9.1.21) was primarily to suggest that, in the absence of any definite information on the nature of active centers at solid surfaces, neglecting the possible nonuniformity of the surface will lead to rate functions that are still *qualitatively* correct. In this sense, we were justified in ignoring nonuniformity in the previous chapters when various general kinetic results concerning catalytic phenomena were discussed. It must be emphasized that the assumptions made in deriving (9.1.21) are very reasonable. Nevertheless, it would appear unwise to adhere too strictly to a formalistic description of kinetics on nonuniform surfaces as long as the physical content of this description remains unspecified, i.e., as long as the exact nature of this nonuniformity remains controversial for lack of direct experimental evidence.

Finally, at the present stage of development of heterogeneous kinetics, even qualitatively correct rate functions are still very useful in both theory and practice. They provide the parameters required to explore various correlations which pave the way to further understanding.

Problem 9.1.1

Consider a two-step catalytic sequence, assume that the second one is rate-determining and derive a rate function for a uniform surface and for a surface that is nonuniform in the sense of Section 9.1.

Problem 9.1.2

Derive an adsorption isotherm, i.e., a relation between the surface and bulk concentrations at a fixed temperature. Consider the simple equilibrium:

$$A \leftrightarrow A_a$$

where A_a is adsorbed A. Assume that the heat of adsorption decreases linearly with surface coverage.

9.2 *Activity and Selectivity*

There are two basic problems concerning catalysts. The first problem is a problem of *activity:* If a given reaction is studied on different catalysts, how does the rate change from catalyst to catalyst? Differences in activity are frequently very large and easy to measure. Whenever possible though, rates should be expressed per unit surface area and not per unit weight or volume of catalyst (see p. 12).

The second question is a more subtle one because it usually involves smaller differences in rates: When a given catalyst is used for various related reactions, how does the rate change from reaction to reaction? This is the problem of *selectivity*. Studies of activity put more emphasis on the catalyst while investigations of selectivity focus attention on the catalyzed reaction itself.

In both cases, it may happen that the rate function has a different form on different catalysts or for different reactions. In this case, the most appropriate kinetic parameter to be used for correlation purposes is the temperature T_r at which the rate has a set value r. This temperature T_r is a far more natural parameter than the rates themselves at a set temperature. Indeed, in kinetic investigations, it is not convenient to measure very fast or very slow rates in a usual apparatus. Therefore, the temperature of kinetic runs is usually adjusted so as to obtain a convenient value of the rate, neither too fast nor too slow. Thus T_r is a natural correlating parameter whereas rates at a fixed temperature may have to be calculated by questionable extrapolations of actual data.

On the other hand, if the form of the rate function remains invariant from catalyst to catalyst or from reaction to reaction, it is generally possible to compare rates by comparing rate constants k in a rate function of the form of the law of mass action. These generally depend on temperature following the law of Arrhenius and the question immediately arises: Are differences in activity or selectivity attributed to changes in activation energy E or in pre-exponential factors A or to simultaneous changes in both?

It is an empirical fact that all three possibilities do occur and representative cases have been summarized in Table 9.2.1 with pertinent details in accompanying tables or figures.

The examples chosen to illustrate changes in activity all pertain to catalysis with metallic films evaporated *in vacuo* so that at least the chemical composition of the catalyst is definitely known. For both activity and selectivity, cases of type A are not unexpected and have their counterpart in homogeneous kinetics where, for instance in the Brönsted relation, changes in rate are attributed to changes in activation energy alone (see p. 172). Cases of the B type are more surprising and they are apparently found only in heterogeneous catalysis. They do not conform to the intuitive idea that a catalyst will be more active or selective if the activation energy is reduced in appropriate fashion.

The compensation effect (type C) is widespread and can be characterized as in the case of homogeneous reactions in solution (see p. 179). In heterogeneous catalysis where the effect was first noticed, the existence of an isokinetic temperature θ is sometimes called the "Theta Rule."

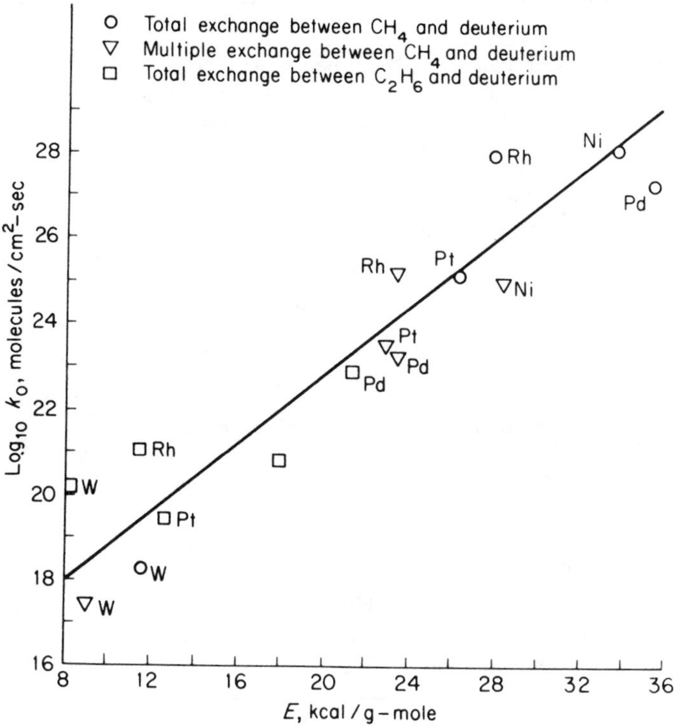

Fig. 9.2.1 Compensation effect for exchange reactions on metals. [Data from J. R. Anderson, and C. Kemball, *Proc. Roy. Soc.* (London), **A223**, 361 (1954) and C. Kemball, *Proc. Roy. Soc.* (London), **A217**, 376 (1953); correlation from M. Boudart, *Chem. Eng. Progress*, **57**, 33 (1961)]

The compensation effect in contact catalysis must be regarded as purely empirical although many explanations for it have been offered. Equally empirical is the classification presented here; it is not possible at the present time to predict whether case A, B or C will occur in a given study of activity or selectivity. The possibility of the three cases must be borne in mind and the case that applies must be clearly recognized before the kinetic parameters (E or A) are used in further correlations based on the properties of the solid

Table 9.2.1

CHANGES IN ACTIVATION ENERGY AND PRE-EXPONENTIAL FACTOR OR COMPENSATION EFFECT

Class	Parameters	Description	Table or Figure
I. Activity			
A	Constant *A* Variable *E*	Exchange between ammonia and deuterium on various metals	Table 9.2.2
B	Constant *E* Variable *A*	Hydrogenation of ethylene on various metals	Table 9.2.3
C	Both *E* and *A* variable with compensation	Exchange between methane or ethane and deuterium on various metals	Fig. 9.2.1
II. Selectivity			
A	Constant *A* Variable *E*	Three-half-order cracking of paraffins on silica-alumina	Table 9.2.4
B	Constant *E* Variable *A*	Hydrogenation of alkyl-aromatics on Raney nickel	Table 9.2.5
C	Both *E* and *A* variable with compensation	First-order cracking of paraffins on silica-alumina	Fig. 9.2.2

Table 9.2.2

ACTIVATION ENERGIES FOR THE EXCHANGE REACTION BETWEEN AMMONIA AND DEUTERIUM*

Metal	*E*, kcal/g-mole
Pt	5.2
Rh	6.7
Pd	8.5
Ni	9.3
W	9.2
Fe	12.5
Cu	13.4
Ag	14.1

*C. Kemball, *Proc. Roy. Soc.* (London) **A214,** 413 (1952).

Table 9.2.3

RATE CONSTANTS k (ARBITRARY UNITS) AT 0°C, PER UNIT SURFACE AREA FOR THE HYDROGENATION OF ETHYLENE (E = 10.7 kcal/g-mole)*

Metal	log k
Rh	4.00
Pd	3.20
Pt	2.36
Ni	1.25
Fe	1.00
W	0
Cr	−0.23
Ta	−0.38

*O. Beeck, *Rev. Mod. Phys.*, **17,** 61 (1945)

Table 9.2.4

CRACKING OF PARAFFINS ON SILICA-ALUMINA, THREE-HALF-ORDER KINETICS*

$k(\text{sec}^{-1}, \text{mm Hg}^{-1/2}, \text{m}^{-2}) = 5 \times 10^2 \exp(-E/RT)$

Paraffin	E, kcal/g-mole
n-butane	26.3
n-pentane	20.7
n-hexane	18.4

*J. L. Franklin and D. E. Nicholson, *J. Phys. Chem*, **60,** 59 (1956)

Table 9.2.5

RATE CONSTANTS AT 170°C (g-mole/minute-g catalyst) FOR THE LIQUID PHASE HYDROGENATION OF AROMATIC HYDROCARBONS ON NICKEL (E = 13 kcal/g-mole)*

Molecules	Rate constants, $k \times 10^3$
Benzene	49
Toluene	38
Cumene	27.5
Ethyl benzene	22.4
p-xylene	12.9
m-xylene	10.4
o-xylene	7.8
Naphthalene	6.7

*J. P. Wauquier and J. C. Jungers, *Bull. Soc. Chim. France*, 1280 (1957).

catalyst or of the reactants. Such correlations are beyond the scope of this text since they concern the use of kinetic data in either theoretical or practical work.

Problem 9.2.1

Find the isokinetic temperature corresponding to the data reported in Fig. 9.2.2.

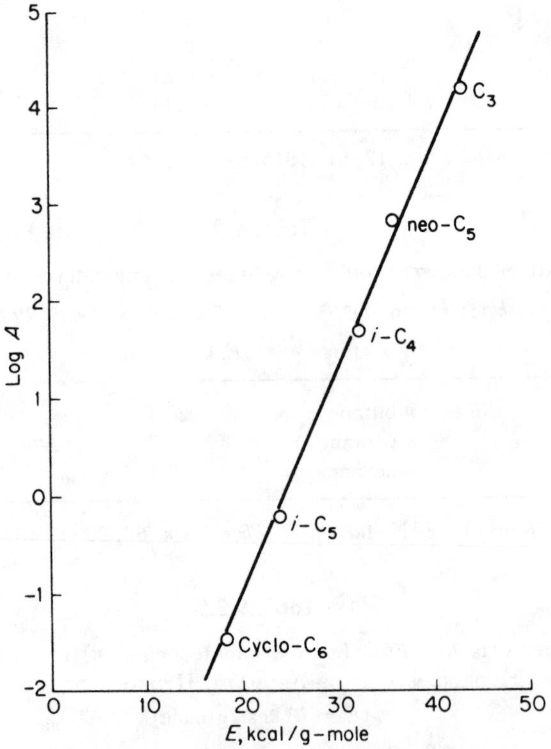

Fig. 9.2.2 Compensation effect for hydrocarbon cracking. [Data from J. F. Franklin and D. E. Nicholson, *J. Phys. Chem.* **60,** 59 (1956), **61,** 814 (1957); correlation from M. Boudart, *Adv. Catalysis,* **9,** 636 (1957)]

9.3 *The Principle of Sabatier*

In any catalytic sequence, the catalyst first forms an "unstable intermediate compound" with one of the reactants. This "intermediate compound" then reacts away and ultimately, at the end of the closed sequence, the catalyst is regenerated. To find the structure and reactivity of the "intermediate compounds" is the goal of research in catalysis.

At solid surfaces, the "unstable intermediate compounds" or "adsorption complexes" perhaps resemble at least *qualitatively* bulk solid compounds that can be isolated, prepared and studied. This idea constitutes the hypothesis put forward by Sabatier.

How can Sabatier's principle be expressed in kinetic terms? A correlation may be expected between the activity of a series of catalysts for a given reaction and the exothermic heat of formation of the intermediate compound. Indeed, following Polanyi's relation (8.1.2), if the heat of formation is too small, the activation barrier for its formation will be too high and the rate of reaction will be small, because the rate-determining step in the catalytic sequence is then the formation step. On the other hand, if the heat of formation is too large, the rate-determining step will be the decomposition of the intermediate compound and this will proceed at a slow rate because of a high activation energy. Therefore, it is expected that the rate of reaction should first increase with heat of formation of the intermediate compound, go through a maximum and then decrease again as the heat of formation becomes too large.

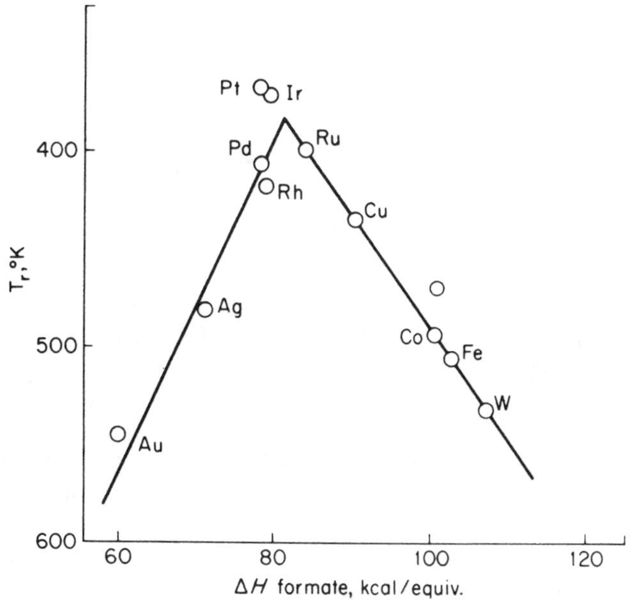

Fig. 9.3.1 Sabatier's principle. Comparison of various catalysts for the formic acid decomposition. [W. J. M., Rootsaert, and W. M. H. Sachtler, *Z. Physik Chem.*, **26,** 16 (1960)]

This behavior is illustrated in Fig. 9.3.1 for the decomposition of formic acid on various metals. Both the ascending and descending branches of the expected activity curve are clearly shown. Since the rate functions are different for different metals, the temperature T_r at which the rate has a set value is used as the correlating kinetic parameter; a high activity (rate) corresponds to a low value of T_r and vice versa.

In this example, the correlating parameter relative to the intermediate compound was the heat of formation of a *bulk* compound, the metal formate. Naturally, it would seem preferable to use the heat of formation of a *surface* compound, i.e., the heat of adsorption. This quantity is used in Figs. 9.3.2

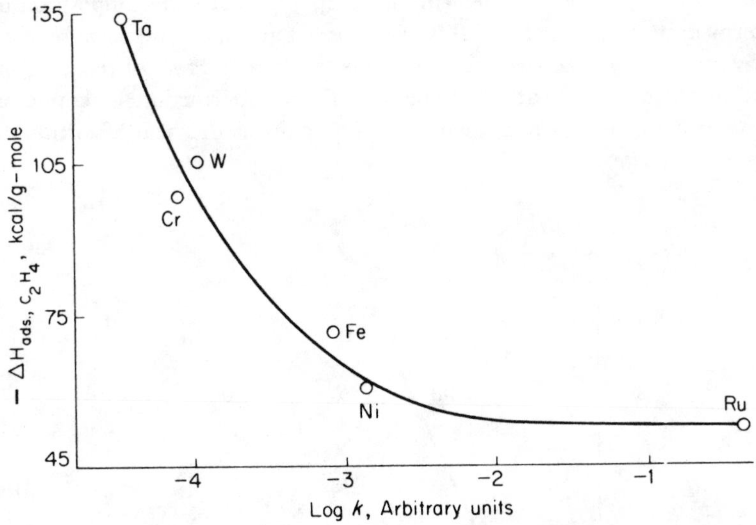

Fig. 9.3.2 Correlation between rate constants for the hydrogenation of ethylene and the heat of adsorption of ethylene. [O. Beeck, *Disc. Faraday Society*, **8**, 118 (1950)]

and 9.3.3, and we see that the rate of hydrogenation of ethylene on various metals goes down as the calorimetric heat of adsorption of both hydrogen and ethylene goes up. On these figures, only the descending branch of the complete curve can be seen. It must be noted that heats of adsorption, like heats of formation of bulk compounds, are still only indirectly related to the heat of formation of the "unstable intermediate compound," i.e., the active center in the catalytic sequence.

In order to look at the problem more quantitatively, let us return to the two-step sequence considered in Section 9.1.:

$$\mathbf{S}_1 + A_1 \rightleftarrows B_1 + \mathbf{S}_2 \qquad (1)$$

$$\mathbf{S}_2 + A_2 \rightleftarrows B_2 + \mathbf{S}_1 \qquad (2)$$

As before, let the first step be exothermic with a positive heat q_1 as befits an adsorption process corresponding to the formation of the unstable intermediate compound $\mathbf{S}_2$. Correspondingly, the second step will be endothermic.

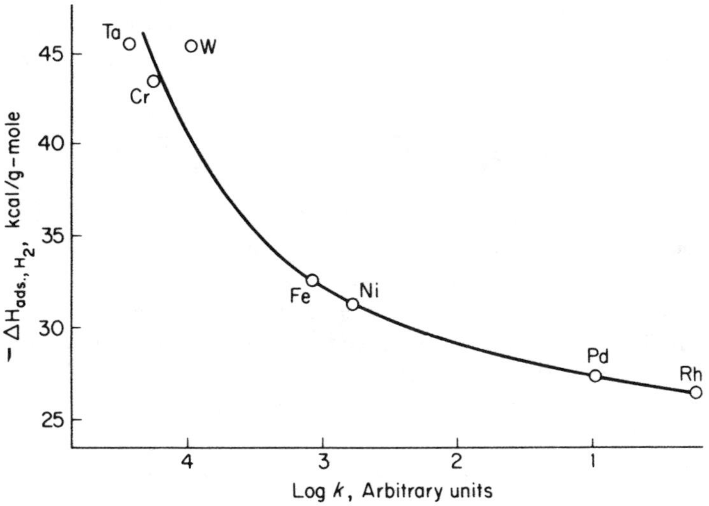

Fig. 9.3.3 Correlation between rate constants for the hydrogenation of ethylene and the heat of adsorption of hydrogen. [O. Beeck, *Disc. Faraday Society*, **8,** 118 (1950)]

Let the overall reaction

$$A_1 + A_2 \rightleftarrows B_1 + B_2$$

take place on the uniform surfaces of various catalysts. The Polanyi relation is expected to apply to the various surfaces as it was applied in Section 9.1 to the various fractions of a nonuniform surface. Let us write the Polanyi relation in the form (8.2.3) for the rate constants of steps (1) and (2):

$$k_1 = C_1 K_1^{\alpha} \tag{9.3.1}$$

$$k_{-1} = C_1 K_1^{-(1-\alpha)} \tag{9.3.2}$$

$$k_2 = C_2 K_2^{-(1-\alpha)} \tag{9.3.3}$$

$$k_{-2} = C_2 K_2^{\alpha} \tag{9.3.4}$$

These relations are analogous to the Brönsted relation (8.2.12).

Further we have:

$$K_1 K_2^{-1} = K \tag{9.3.5}$$

i.e., the ratio of the equilibrium constants of steps (1) and (2) is equal to the equilibrium constant K of the overall reaction. It can now be shown that if α is equal to $\frac{1}{2}$, the most active catalyst will be the one for which $(S_1) = (S_2)$, i.e., the catalyst for which the fraction of the surface covered is equal to $\frac{1}{2}$.

Indeed, substitution of (9.3.1) to (9.3.5) into (9.1.22) gives for $\alpha = \frac{1}{2}$:

$$r = \frac{C_1 C_2 [K^{1/2}(A_1)(A_2) - K^{-1/2}(B_1)(B_2)]}{C_1 K_1^{1/2}(A_1) + C_1 K_1^{-1/2}(B_1) + C_2 K_2^{1/2}(A_2) + C_2 K_2^{-1/2}(B_2)} \tag{9.3.6}$$

Now C_1 and C_2 do not depend on the nature of the catalyst since, in the derivation of the Brönsted relations (9.3.1) to (9.3.4), it is assumed (see p. 172) that entropies of activation and of reaction do not change from one member of the family of reactions to the next. Therefore, the numerator of (9.3.6) is the same on all catalysts and the rate will be maximum on the catalyst for which the denominator D of (9.3.6) has its minimum value. With $x = K_1^{1/2}$ let us rewrite the denominator D in the form:

$$[C_1(A_1) + C_2 K^{-1/2}(B_2)]x + [C_1(B_1) + C_2 K^{1/2}(A_2)]x^{-1}$$

The condition $(dD/dx) = 0$ gives:

$$C_1(A_1) + C_2 K^{-1/2}(B_2) = [C_1(B_1) + C_2 K^{1/2}(A_2)]x^{-2} \tag{9.3.7}$$

Multiplying both sides of (9.3.7) by $K_1^{1/2}$, we obtain the condition for maximum rate:

$$k_1(A_1) + k_{-2}(B_2) = k_{-1}(B_1) + k_2(A_2) \tag{9.3.8}$$

But, according to (9.1.14) and (9.1.15), this condition is equivalent to

$$(S_1) = (S_2) \tag{9.3.9}$$

Thus, if $\alpha = \frac{1}{2}$, the surface coverage is equal to $\frac{1}{2}$. From qualitative arguments similar to those used above concerning the heat of formation of the active intermediate compound, it seems reasonable that the best catalyst is one for which the surface coverage by this compound is neither too large nor too small. But that the surface coverage should be $\frac{1}{2}$ can be demonstrated only for the case $\alpha = \frac{1}{2}$.

Although the theorem appears to have limited applicability, it has a corollary which is of great practical significance. It is of course true that a substance that acts like a catalyst for a reaction in the forward direction, must also be a catalyst for the same reaction in the reverse direction. But it is not generally true that the most active catalyst for a reaction from left to right is also the most active catalyst for the same reaction in the reverse direction.

It is sufficient to show this for one particular case, the one just considered for $\alpha = \frac{1}{2}$.

Indeed, for the two-step catalytic sequence, the condition (9.3.8) for maximum rate becomes

$$k_1(A_1) = k_2(A_2) \tag{9.3.10}$$

when the reaction is carried out far from equilibrium from left to right. But, for the reaction carried out from right to left far from equilibrium, the condition must be:

$$k_{-1}(B_1) = k_{-2}(B_2) \tag{9.3.11}$$

Clearly, (9.3.10) and (9.3.11) will not generally be satisfied simultaneously on a given catalyst and therefore, far from equilibrium, the most active catalyst will not necessarily be the same for both the forward and the reverse reactions.

This limitation is well known empirically in heterogeneous catalysis where frequently, for reasons of convenience, a reaction is studied on a series of catalysts, far from equilibrium in a direction opposed to that of actual interest. For example, a decomposition may be easier to study than the corresponding synthesis. While such studies may be very rewarding, the limitation concerning the most active catalyst must be kept in mind.

Problem 9.3.1

Find the condition defining the most active catalyst for the conditions leading to the rate function (9.1.21). Assume $\alpha = \frac{1}{2}$. Discuss the striking result obtained.

9.4 A Kinetic Comparison Between Homogeneous and Heterogeneous Catalysis

Contact catalysis offers many practical advantages over homogeneous catalysis. Thus, a solid catalyst can be separated from the reaction medium at the end of reaction. Many solid catalysts are versatile, in the sense that they will exhibit good activity in a large number of situations. They usually are resistant to heat and, while they are usually not too selective, this disadvantage is counterbalanced by the complementary virtue of not being too susceptible to inhibition or adulteration by foreign impurities.

But these advantages are not the only reason for the success of solid catalysts, as compared to soluble catalysts, in the present stage of development of chemistry. It appears that a decisive advantage of solid surfaces is that they exhibit nonuniformity in the kinetic sense discussed in this chapter. Although a discussion of the nonuniformity itself is a subject beyond the scope of chemical kinetics, it is pertinent to conclude this chapter with a kinetic comparison between a catalyst presenting a nonuniform surface and one that presents a uniform surface. But the latter could equally be a soluble catalyst so that, actually, we shall compare, from a kinetic standpoint, homogeneous and heterogeneous catalysis, if nonuniformity is considered as a normal property of solid catalysts.

Consider the reaction A $\longrightarrow$ B taking place on a uniform surface at the rate r_u and on a nonuniform surface at the rate r_n.

Assume that adsorption of A is rate-determining and that adsorbed B is the most abundant species at the surface. Then:

$$r_u = k(\text{A})(1 - \theta) = \frac{k(\text{A})}{1 + K(\text{B})} \tag{9.4.1}$$

where θ is the fraction of surface covered with B and K is the adsorption equilibrium constant of B.

Assume further that, in Brönsted fashion (p. 175):

$$k = GK^{\alpha} = GK^{1/2} \tag{9.4.2}$$

with, for simplicity, $\alpha = \frac{1}{2}$. The quantity G is a constant.

Substitution of (9.4.2) into (9.4.1) gives:

$$r_u = \frac{GK^{1/2}(\text{A})}{1 + K(\text{B})} \tag{9.4.3}$$

The optimum surface will be one for which

$$\frac{\partial r_u}{\partial K} = 0 \tag{9.4.4}$$

Condition (9.4.4) gives for the optimum catalyst with a uniform surface:

$$K = (\text{B})^{-1} \qquad \text{or} \qquad \theta = \tfrac{1}{2}$$

The optimum rate is then:

$$r_{u,\max} = \left(\frac{1}{2}\right)G(\text{A})(\text{B})^{-1/2} \tag{9.4.5}$$

On a nonuniform surface, Eq. (9.1.21) with $\alpha = \frac{1}{2}$ and with (9.4.2) gives:

$$r_n = \frac{\pi}{f}\, G(\text{A})(\text{B})^{-1/2} \tag{9.4.6}$$

Comparing (9.4.5) and (9.4.6), we obtain the interesting result:

$$\boxed{\frac{r_{u,\max}}{r_n} = \frac{f}{2\pi}} \tag{9.4.7}$$

Thus, the ratio of rates on the optimum uniform surface and on the nonuniform surface is determined by the parameter of nonuniformity f. Similarly, let us find the ratio r_i/r_n where r_i is the rate on sites characterized by a constant K_i and a heat of adsorption q_i differing by an amount $\Delta q = q_i - q_{\max}$ from the heat $q_{\max}$ of sites with a maximum rate.

We have:

$$K_{\max} = (\text{B})^{-1} = K_0\, e^{q_{\max}/RT} \tag{9.4.8}$$

$$K_i = K_0\, e^{q_i/RT} \tag{9.4.9}$$

Dividing (9.4.8) and (9.4.9) side by side, we get:

$$K_i = (\text{B})^{-1} \exp\left(\frac{\Delta q}{RT}\right) \tag{9.4.10}$$

But, from (9.4.3), we have:

$$r_i = \frac{GK_i^{1/2}(\text{A})}{1 + K_i(\text{B})} \tag{9.4.11}$$

From (9.4.6), (9.4.10), and (9.4.11), we get:

$$\boxed{\frac{r_i}{r_n} = \frac{f \exp\left(\dfrac{\Delta q}{2RT}\right)}{\pi\left[1 + \exp\left(\dfrac{\Delta q}{RT}\right)\right]}} \tag{9.4.12}$$

Similarly, from (9.4.5), (9.4.10), and (9.4.11), we get

$$\boxed{\frac{r_{u,\max}}{r_i} = \frac{1 + \exp\left(\frac{\Delta q}{RT}\right)}{2 \exp\left(\frac{\Delta q}{2RT}\right)}} \tag{9.4.13}$$

An interesting consequence from (9.4.12) can now be obtained: Let us calculate Δq for which $r_i = r_n$:

$$f \exp\left(\frac{\Delta q}{2RT}\right) = \pi\left[1 + \exp\left(\frac{\Delta q}{RT}\right)\right] \tag{9.4.14}$$

If $\exp(-\Delta q/RT) \ll 1$, (9.4.14) gives more simply

$$\boxed{\Delta q = 2RT \ln\left(\frac{f}{\pi}\right)} \tag{9.4.15}$$

The physical content of these relations is best explored by means of a numerical example. Let us imagine a catalyst with a linear variation of heats of

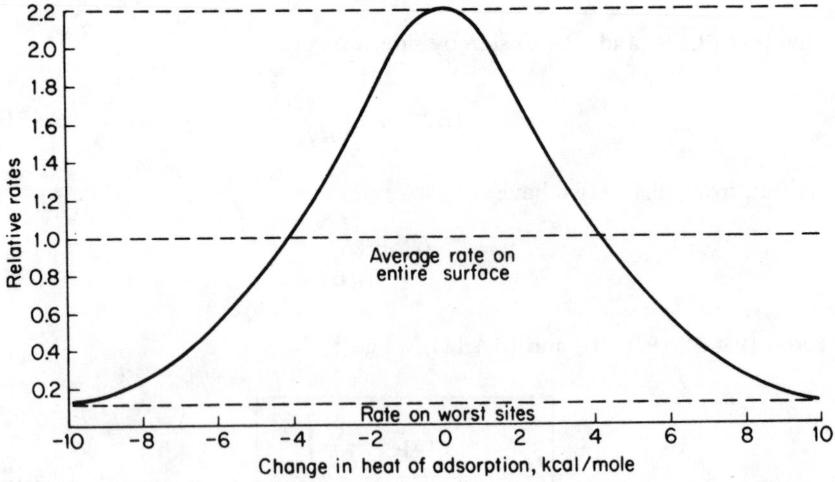

Fig. 9.4.1 Behavior of a nonuniform surface with a linear spread of heats of adsorption equal to 20 kcal and a value of α equal to $\frac{1}{2}$, at 450°C.

adsorption with surface coverage, the variation amounting to 20 kcal/g-mole, or $\Delta q = \pm 10$ kcal/g-mole. At 450°C, the parameter of nonuniformity f is then equal to 13.8. Hence from (9.4.7), the rate on the optimum uniform surface $r_{u,\max}$ would be only 2.2 times that on the nonuniform surface. From (9.4.13), we calculate that the rate r_i at both edges of the distribution $\Delta q = \pm 10$ kcal/g-mole, is only 17.4 times smaller than the rate $r_{u,\max}$ on the optimum sites. From (9.4.15), we calculate that if the heat of adsorption on the uniform surface deviated by as little as $\Delta q = 4.3$ kcal/g-mole, from the optimum value, the rate would fall to the value it has on the nonuniform surface.

These results, which are shown on Fig. 9.4.1, lead to the following conclusions. If the surface is broadly nonuniform as established experimentally in a large number of cases, it is incorrect to think that only a few sites on this nonuniform surface do all the work. Actually, even the poorest sites, at the edges of the distribution, still contribute significantly to the rate. Moreover, it appears that a nonuniform surface is the key to the success of a solid catalyst. Indeed, even if by great skill, it were possible to prepare an optimum catalyst consisting of uniform active centers, this optimum catalyst would cease to be optimum if either the temperature or the conversion were changed, since the condition for optimum activity derived from (9.4.4) was $K = (B)^{-1}$. The nature of the nonuniformity of the surface is then clearly one of the central problems of heterogeneous catalysis.

BIBLIOGRAPHY

9.1 The kinetic treatment of a nonuniform surface, as presented here, is taken from a paper by M. I. Temkin, *Zhur. Fiz. Khim.*, **31**, 1 (1957). Temkin has presented a generalization of these results in *Dok. Akad. Nauk SSSR*, **161**, 160 (1965).

9.2 Other pertinent details on correlations in heterogeneous catalysis have been summarized by the author in *Chem. Eng. Progress*, **57**, No. 8, 33 (1961).

9.3 A very valuable introduction to contact catalysis is found in Paul H. Emmett's Priestley Lecture, "New Approaches to the Study of Catalysis," Pennsylvania State University, University Park, Pennsylvania, 1962.

9.4 The results of this section are those of the school of Temkin, as found in a book by S. L. Kiperman, *Introduction to the Kinetics of Heterogeneous Catalytic Reactions*, Moscow, 1964. Similar ideas have been presented many times by Taylor and his school.

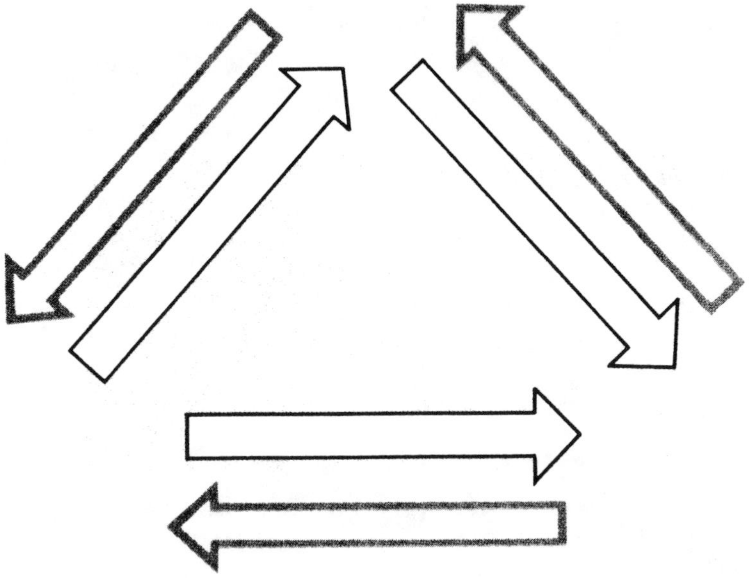

Analysis of Reaction Networks

10

All too frequently, a large number of single reactions takes place in succession or parallel in a reacting system. It may be hard or impossible to isolate any single reaction for separate study and dissection into its elementary steps. Rather the *network of reactions* must be analyzed as a whole. This is a problem of considerable difficulty and every case taxes the ingenuity of the kineticist.

In the standard textbooks, it is customary to include analytical solutions of the differential rate equations for a handful of cases, largely of academic interest. A study of this material can be rewarding in that it reveals the cumbersome nature of available solutions and the inherent difficulty of extracting information on rate constants and reaction orders from the usual kinetic data giving composition as a function of time.

The need for analytical solutions has now largely disappeared thanks to the availability of analog and digital computing machines. But the difficulty of extracting *meaningful* kinetic information remains the same or perhaps has even increased since much more complicated networks are now amenable to numerical analysis.

It is therefore particularly urgent to supplement kinetic data with theoretical insight and experimental techniques that may guide the experimentalist in gathering an informative set of data. Two such approaches will be discussed in this chapter.

10.1 The Kinetic Tracer Technique of Neiman

The first question to resolve before attempting the kinetic analysis of a network is the proper ordering of reactions in the network. Chemical intuition is of great assistance but is rarely sufficient.

Consider, for example, the catalytic oxidation of naphthalene N. On certain solid catalysts, phthalic anhydride P is a major product while on other catalysts important quantities of naphthoquinone Q are produced. In all cases, a certain quantity of maleic anhydride M also appears as well as gaseous products G of complete oxidation, CO, CO_2, and H_2O.

A possible network would be the following:

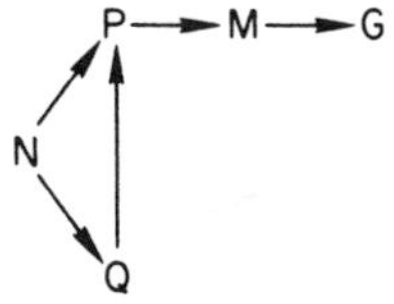

It is important to verify beforehand the plausibility of this network. For instance, is P truly coming from both N and Q? Is M the sole precursor of G or is G also coming directly from Q, or perhaps also from N and P?

Isotopic labeling of selected components of the system frequently provides an answer to such questions. The technique possesses many different aspects adapted to individual cases. As an example, let us cite the many investigations of Melvin Calvin and collaborators over a period of twenty years. In these studies, systematic use was made of carbon dioxide labeled with ^{14}C to elucidate the early steps of photosynthesis and the network of reactions involved in the assimilation of carbon dioxide by green plants. An important variation of the technique consists in the use of tracers in a kinetic sense. Among the several possibilities of this kind proposed by Neiman, one deserves special attention.

Indeed, in many networks, as illustrated above, it is possible to isolate a segment

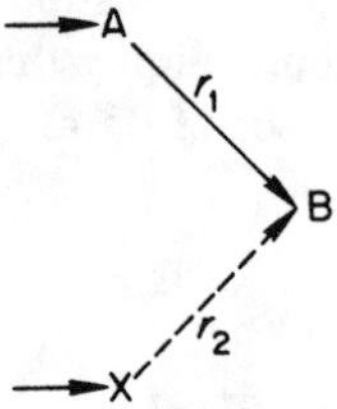

for which the question to be asked is: Does B come only from A or is X perhaps also a precursor of B? Let us assume that the rate of the reaction A → B is r_1 while r_2 is the rate of the questionable reaction X → B. Let us introduce in the reacting system a small quantity of a molecular species A, in case it is not one of the original reactants. A very small quantity of A is tagged with, say, a radioactive isotope. The tagged molecule is denoted A*. Let us follow in time the specific activity $\alpha = (A^*)/(A)$ of component A and the specific

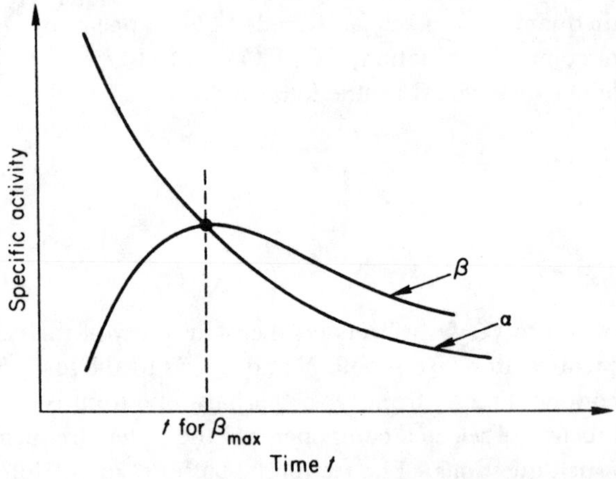

Fig. 10.1.1 Neiman's technique to verify that a given species in a network is the sole precursor of another one.

activity $\beta = (B^*)/(B)$ of component B. Clearly, α decreases monotonically with time (Fig. 10.1.1). But what happens to β? It will increase at first, pass through a maximum, then decrease:

$$\frac{d\beta}{dt} = \frac{d\left[\frac{(B^*)}{(B)}\right]}{dt} = \frac{1}{(B)}\frac{d(B^*)}{dt} - \frac{(B^*)}{(B)^2}\frac{d(B)}{dt} \qquad (10.1.1)$$

But it is evident that:

$$\frac{d(B)}{dt} = r_1 + r_2 \tag{10.1.2}$$

$$\frac{d(B^*)}{dt} = \alpha r_1 \tag{10.1.3}$$

Substitution of (10.1.2) and (10.1.3) into (10.1.1) gives:

$$\frac{d\beta}{dt} = \frac{1}{(B)}[\alpha r_1 - \beta(r_1 + r_2)] \tag{10.1.4}$$

Equation (10.1.4) shows that the maximum value in β is reached when:

$$\frac{\alpha}{\beta} = \frac{r_1 + r_2}{r_1} \tag{10.1.5}$$

Thus, if for instance A is the sole precursor of B, $r_2 = 0$ and the values of α and β are equal at the maximum value in β, as shown in Fig. 10.1.1.

Another problem, in which the kinetic use of labeled molecules following Neiman can be useful, arises in any network where the segment:

$$A \xrightarrow{r_1} B \xrightarrow{r_2} C$$

can be isolated. Then the rate r of accumulation of B is:

$$r = r_1 - r_2$$

If a small quantity of B labeled with B* is introduced in the system and if the specific activity β of B is followed with time, it is possible to determine r_1 and r_2 separately from a measurement of r. Indeed, the same equation (10.1.1) holds. But now:

$$\frac{d(B)}{dt} = r_1 - r_2 \tag{10.1.6}$$

and

$$\frac{d(B^*)}{dt} = -\beta r_2 \tag{10.1.7}$$

Substitution of (10.1.6) and (10.1.7) into (10.1.1) gives:

$$\frac{d\beta}{dt} = -\beta \frac{r_1}{(\mathrm{B})}$$

Consequently:

$$r_1 = -(\mathrm{B}) \frac{d(\ln\beta)}{dt} \tag{10.1.8}$$

Therefore measurements of both (B) and β as a function of time allow us to calculate r_1 and therefore also r_2 separately.

Finally, and almost patently, in a network with a segment

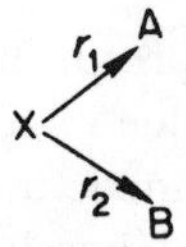

the ratio of the rate r_1 and r_2 can be measured by following the relative rate of appearance of a tracer into components A and B:

$$\frac{r_1}{r_2} = \frac{d\alpha}{d\beta} \tag{10.1.9}$$

Problem 10.1.1

Show that Eq. (10.1.5) holds without modification for a network with a segment:

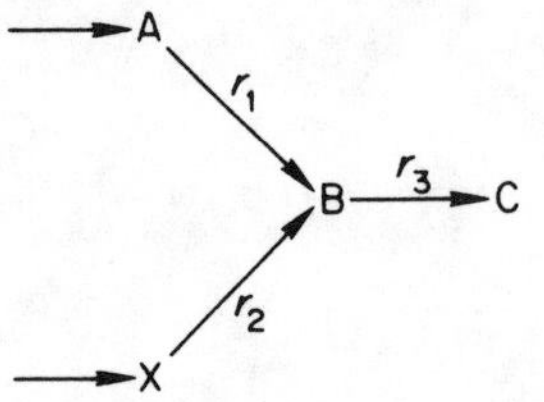

10.2 Kinetic Behavior of a Network of First-order Reversible Reactions: The Method of Wei and Prater

Consider n components A_i ($i = 1,2, \ldots i,j, \ldots n$) interconnected by means of reversible first-order reactions. The rate constant of the reaction where A_i is transformed into A_j will be called k_{ij}. Some of the pairs of rate constants k_{ji} and k_{ij} may be equal to zero, but at equilibrium the system contains finite amounts a_i^* of all components. The composition of the system at any time is denoted by mole fractions a_i of all components A_i.

In order to determine the kinetic behavior of the system, it is necessary to determine the value of all rate constants which are generally unknown.

An elegant method for determining the rate constants is that developed by Wei and Prater. Its interest lies in the fact that it dictates a mode of gathering kinetic data that is not obvious, yet presents many advantages. The method will be explained for a three-component system, but generalization to a multicomponent system follows immediately from the matrix notation, which will be introduced because of the useful geometrical interpretation that it affords.

If we select a Cartesian coordinate system and use its axes to represent composition, each point representing the system at any time can be represented by a vector α with components a_1, a_2, a_3. Because $a_1 + a_2 + a_3 = 1$, the extremity of the vector will be confined within a triangle which can be used conveniently to represent the evolution of a system of any arbitrary initial composition. The trajectory in the reaction triangle is called a *reaction path*. In general, as shown in Fig. 10.2.1, the reaction path is curved.

Now the system of differential equations representing the behavior of the system:

$$\frac{da_1}{dt} = -(k_{12} + k_{13})a_1 + k_{21}\,a_2 + k_{31}\,a_3$$

$$\frac{da_2}{dt} = k_{12}\,a_1 - (k_{23} + k_{21})a_2 + k_{32}\,a_3$$

$$\frac{da_3}{dt} = k_{13}\,a_1 + k_{23}\,a_2 - (k_{31} + k_{32})a_3$$

can be written in matrix notation as:

$$\frac{d\boldsymbol{\alpha}}{dt} = \mathbf{K}\boldsymbol{\alpha} \tag{10.2.1}$$

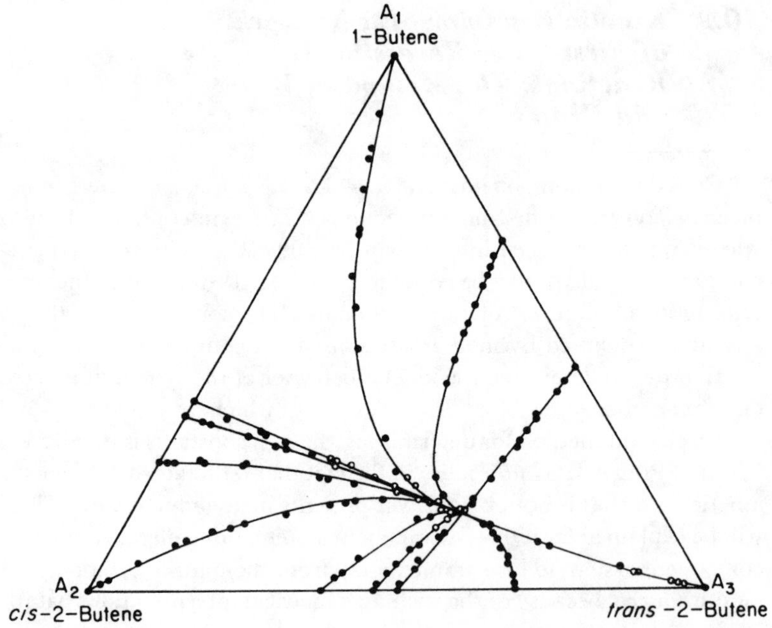

Fig. 10.2.1 Comparison of calculated reaction paths with experimentally observed compositions for butene isomerization. The points are observed composition and the solid lines are calculated reaction paths. [From James Wei and Charles D. Prater, *Adv. in Cat.*, **13**, 256 (1962). Reproduced with permission.]

where **K** is a square matrix

$$\mathbf{K} = \begin{bmatrix} -(k_{12} + k_{13}) & k_{21} & k_{31} \\ k_{12} & -(k_{23} + k_{21}) & k_{32} \\ k_{13} & k_{23} & -(k_{31} + k_{32}) \end{bmatrix} \tag{10.2.2}$$

and $\boldsymbol{\alpha}$ is the column vector

$$\boldsymbol{\alpha} = \begin{bmatrix} a_1 \\ a_2 \\ a_3 \end{bmatrix} \tag{10.2.3}$$

Let us now assume that the matrix **K** possesses three independent *char-*

acteristic vectors $\mathbf{x}_0$, $\mathbf{x}_1$, and $\mathbf{x}_2$, to which are associated three *characteristic non-positive roots* $-\lambda_0$, $-\lambda_1$, and $-\lambda_2$, such that

$$\mathbf{K}\mathbf{x}_i = -\lambda_i \mathbf{x}_i \qquad (i = 0,1,2) \tag{10.2.4}$$

The components of each vector $\mathbf{x}_i$, whose length so far is arbitrary, are x_{1i}, x_{2i}, x_{3i}. If we form the matrix:

$$\mathbf{X} = \begin{bmatrix} x_{10} & x_{11} & x_{12} \\ x_{20} & x_{21} & x_{22} \\ x_{30} & x_{31} & x_{32} \end{bmatrix} \tag{10.2.5}$$

and the diagonal matrix

$$\mathbf{\Lambda} = (-)\begin{bmatrix} \lambda_0 & 0 & 0 \\ 0 & \lambda_1 & 0 \\ 0 & 0 & \lambda_2 \end{bmatrix} \tag{10.2.6}$$

the three equations (10.2.4) can be written succinctly as:

$$\mathbf{K}\mathbf{X} = \mathbf{X}\mathbf{\Lambda} \tag{10.2.7}$$

Post-multiplying both sides of (10.2.7) by $\mathbf{X}^{-1}$, the inverse of the matrix $\mathbf{X}$, we get:

$$\boxed{\mathbf{K} = \mathbf{X}\mathbf{\Lambda}\mathbf{X}^{-1}} \tag{10.2.8}$$

Therefore the problem of calculating the rate constants consists in determining the matrix $\mathbf{X}$ and its inverse as well as the characteristic roots. How can this be done?

Now, let us use the three directions (generally nonorthogonal) of the characteristic vectors as a new system of coordinates to represent the reacting system. The vectors $\mathbf{x}_i$ will be used as unit vectors. The vector $\boldsymbol{\alpha}$ can now be represented in this new system of coordinates by means of the transformation:

$$\begin{aligned} a_1 &= b_0 x_{10} + b_1 x_{11} + b_2 x_{12} \\ a_2 &= b_0 x_{20} + b_1 x_{21} + b_2 x_{22} \\ a_3 &= b_0 x_{30} + b_1 x_{31} + b_2 x_{32} \end{aligned} \tag{10.2.9}$$

which can be written as

$$\boldsymbol{\alpha} = b_0\mathbf{x}_0 + b_1\mathbf{x}_1 + b_2\mathbf{x}_2 \tag{10.2.10}$$

or

$$\boxed{\boldsymbol{\alpha} = \mathbf{X}\boldsymbol{\beta}} \tag{10.2.11}$$

where $\boldsymbol{\beta}$ is a new vector

$$\boldsymbol{\beta} = \begin{bmatrix} b_0 \\ b_1 \\ b_2 \end{bmatrix} \tag{10.2.12}$$

whose kinetic behavior is exceedingly simple. Indeed, substitution of (10.2.11) into (10.2.1) yields:

$$\mathbf{X}\frac{d\boldsymbol{\beta}}{dt} = \mathbf{K}\mathbf{X}\boldsymbol{\beta} \tag{10.2.13}$$

Pre-multiplying both sides of (10.2.13) by $\mathbf{X}^{-1}$ gives:

$$\frac{d\boldsymbol{\beta}}{dt} = \mathbf{X}^{-1}\mathbf{K}\mathbf{X}\boldsymbol{\beta} \tag{10.2.14}$$

since by definition, the product $\mathbf{X}^{-1}\mathbf{X}$ is the unit matrix which never needs to be written explicitly. On the other hand, pre-multiplication of both sides of (10.2.7) by $\mathbf{X}^{-1}$ gives:

$$\boldsymbol{\Lambda} = \mathbf{X}^{-1}\mathbf{K}\mathbf{X} \tag{10.2.15}$$

Substitution of (10.2.15) into (10.2.14) gives:

$$\frac{d\boldsymbol{\beta}}{dt} = \boldsymbol{\Lambda}\boldsymbol{\beta} \tag{10.2.16}$$

or explicitly:

$$\begin{aligned} \frac{db_0}{dt} &= -\lambda_0 b_0 \\ \frac{db_1}{dt} &= -\lambda_1 b_1 \\ \frac{db_2}{dt} &= -\lambda_2 b_2 \end{aligned} \tag{10.2.17}$$

The first characteristic vector and root can be found immediately. Indeed at equilibrium:

$$\frac{d\boldsymbol{\alpha}}{dt} = 0 \tag{10.2.18}$$

From (10.2.17) and (10.2.1) we get:

$$\mathbf{K}\boldsymbol{\alpha}^* = 0 = 0\boldsymbol{\alpha}^* \tag{10.2.19}$$

Comparing (10.2.19) with (10.2.4) we see that the equilibrium vector $\boldsymbol{\alpha}^*$ is a characteristic vector with $\lambda_0 = 0$.

Thus:

$$\mathbf{x}_0 = \boldsymbol{\alpha}^* \tag{10.2.20}$$

and the sum of the components of $\mathbf{x}_0$, namely $a_1^* + a_2^* + a_3^*$, is equal to unity.

The length of the first characteristic vector being determined, let us now fix the length of the others. This will be done by means of the equations

$$\begin{aligned} \mathbf{x}_0 + \mathbf{x}_1 &= \boldsymbol{\alpha}_{x_1}(0) \\ \mathbf{x}_0 + \mathbf{x}_2 &= \boldsymbol{\alpha}_{x_2}(0) \end{aligned} \tag{10.2.21}$$

The vectors $\boldsymbol{\alpha}_{x_1}(0)$ and $\boldsymbol{\alpha}_{x_2}(0)$ represent initial compositions on the sides of the reaction triangle. From the relations (10.2.21), it is clear that the sum of the components of $\mathbf{x}_1$ is equal to zero. The same applies to $\mathbf{x}_2$. Indeed, the components of $\boldsymbol{\alpha}_{x_1}(0)$ add up to unity just as the components of $\mathbf{x}_0$.

The important property of $\boldsymbol{\alpha}_{x_1}(0)$ and $\boldsymbol{\alpha}_{x_2}(0)$ is that these vectors define initial compositions which will decay to equilibrium following *straight-line reaction paths*.

Indeed, keeping in mind that $\lambda_0 = 0$, Eqs. (10.2.17) can be integrated:

$$\begin{aligned} b_0 &= b_0^{(0)} \\ b_1 &= b_1^{(0)} e^{-\lambda_1 t} \\ b_2 &= b_2^{(0)} e^{-\lambda_2 t} \end{aligned} \tag{10.2.22}$$

Substitution of (10.2.22) into (10.2.10) yields:

$$\boldsymbol{\alpha} = b_0^{(0)}\mathbf{x}_0 + b_1^{(0)}e^{-\lambda_1 t}\mathbf{x}_1 + b_2^{(0)}e^{-\lambda_2 t}\mathbf{x}_2 \tag{10.2.23}$$

which gives for an initial system ($t = 0$):

$$\boldsymbol{\alpha}(0) = b_0^{(0)}\mathbf{x}_0 + b_1^{(0)}\mathbf{x}_1 + b_2^{(0)}\mathbf{x}_2 \tag{10.2.24}$$

Comparison of (10.2.24) with (10.2.21) shows that for the initial vectors $\boldsymbol{\alpha}_{x_1}(0)$ and $\boldsymbol{\alpha}_{x_2}(0)$, we must have respectively

$$b_0^{(0)} = 1, \quad b_1^{(0)} = 1, \quad b_2^{(0)} = 0$$

and

$$b_0^{(0)} = 1, \quad b_1^{(0)} = 0, \quad b_2^{(0)} = 1$$

Therefore:

$$\boldsymbol{\alpha}_{x_1}(t) = \mathbf{x}_0 + e^{-\lambda_1 t}\mathbf{x}_1 \qquad (10.2.25)$$

$$\boldsymbol{\alpha}_{x_2}(t) = \mathbf{x}_0 + e^{-\lambda_2 t}\mathbf{x}_2$$

The extremities of these vectors will therefore be straight lines in the reaction triangle, connecting the initial compositions defined by $\boldsymbol{\alpha}_{x_1}(0)$ and $\boldsymbol{\alpha}_{x_2}(0)$ with the equilibrium point defined by $\mathbf{x}_0$.

The existence of these straight-line reaction paths that are accessible to experiment suggests a systematic method of approach to the problem of calculating rate constants for a reaction network.

10.3 Determination of Straight-line Reaction Paths

The characteristic vectors of the rate-constant matrix **K** are unfortunately not orthogonal to each other. The reason for this is that the matrix **K** is not generally symmetric. If it could be made symmetric by a so-called similarity transformation of the type $\bar{\mathbf{K}} = \mathbf{P}^{-1}\mathbf{K}\mathbf{P}$ where **P** is any nonsingular matrix of the same size as **K**, the new matrix $\bar{\mathbf{K}}$ would possess a set of *orthogonal characteristic vectors* $\mathbf{x}_i$ $(i = 1,2, \ldots i, \ldots n)$ with the same characteristic roots λ_i as the original matrix **K**.

Now, the principle of microscopic reversibility (see Section 3.5) provides us with an easy means to find such a transformation. Indeed, let us form the diagonal matrix:

$$\mathbf{D} = \begin{bmatrix} a_1^* & 0 & 0 \\ 0 & a_2^* & 0 \\ 0 & 0 & a_3^* \end{bmatrix} \qquad (10.3.1)$$

It is easy to verify that the matrix **KD** is symmetric. From (10.3.1) and (10.2.2), we get

$$\mathbf{KD} = \begin{bmatrix} -(k_{12} + k_{13})a_1^* & k_{21}a_2^* & k_{31}a_3^* \\ k_{12}a_1^* & -(k_{23} + k_{21})a_2^* & k_{32}a_3^* \\ k_{13}a_1^* & k_{23}a_2^* & -(k_{31} + k_{32})a_3^* \end{bmatrix} \qquad (10.3.2)$$

Since $k_{ij}a_i{}^* = k_{ji}a_j{}^*$ according to the principle of microscopic reversibility, **KD** does not change when its rows are exchanged with its columns:

$$\mathbf{KD} = (\mathbf{KD})^T \tag{10.3.3}$$

That is, the matrix is equal to its transpose T and it is symmetric.

If that is the case, the matrix

$$\bar{\mathbf{K}} = \mathbf{D}^{-1/2}\mathbf{K}\mathbf{D}^{1/2} \tag{10.3.4}$$

is also symmetric. Indeed:

$$\bar{\mathbf{K}}^T = (\mathbf{D}^{-1/2}\mathbf{K}\mathbf{D}^{1/2})^T = \mathbf{D}^{1/2}\mathbf{K}^T\mathbf{D}^{-1/2} \tag{10.3.5}$$

where the rule has been used that the transpose (or the inverse) of a product of matrices is equal to the product of the transpose (or the inverse) of the individual matrices taken in reverse order. It must also be noted that a diagonal matrix is of course symmetric. The two diagonal matrices $\mathbf{D}^{1/2}$ and $\mathbf{D}^{-1/2}$ are respectively:

$$\mathbf{D}^{1/2} = \begin{bmatrix} \sqrt{a_1{}^*} & 0 & 0 \\ 0 & \sqrt{a_2{}^*} & 0 \\ 0 & 0 & \sqrt{a_3{}^*} \end{bmatrix} \tag{10.3.6}$$

and

$$\mathbf{D}^{-1/2} = \begin{bmatrix} 1/\sqrt{a_1{}^*} & 0 & 0 \\ 0 & 1/\sqrt{a_2{}^*} & 0 \\ 0 & 0 & 1/\sqrt{a_3{}^*} \end{bmatrix} \tag{10.3.7}$$

Equation (10.3.3) can be rewritten as

$$\mathbf{KD} = \mathbf{DK}^T \tag{10.3.8}$$

Therefore, pre-multiplication of both sides of (10.3.8) by $\mathbf{D}^{-1}$ gives:

$$\mathbf{K}^T = \mathbf{D}^{-1}\mathbf{KD} \tag{10.3.9}$$

Substitution of (10.3.9) into (10.3.5) gives:

$$\bar{\mathbf{K}}^T = \mathbf{D}^{1/2}\mathbf{D}^{-1}\mathbf{KD}\mathbf{D}^{-1/2} = \mathbf{D}^{-1/2}\mathbf{K}\mathbf{D}^{1/2} = \bar{\mathbf{K}} \tag{10.3.10}$$

Thus the matrix $\bar{\mathbf{K}}$ is symmetric. Since it has the same characteristic roots as $\mathbf{K}$, we can write:

$$\bar{\mathbf{K}}\bar{\mathbf{x}}_i = -\lambda_i\bar{\mathbf{x}}_i \qquad (i = 1,2, \ldots i, \ldots n) \tag{10.3.11}$$

Comparing (10.3.11) with (10.2.4) and using (10.3.4), we verify easily that

$$\boxed{\bar{\mathbf{x}}_i = \mathbf{D}^{-1/2}\mathbf{x}_i} \tag{10.3.12}$$

Consequently, the matrix $\bar{\mathbf{X}}$ of the orthogonal characteristic vectors of $\bar{\mathbf{K}}$ is related to the matrix $\mathbf{X}$ of the characteristic vectors of $\mathbf{K}$ by means of the relation:

$$\bar{\mathbf{X}} = \mathbf{D}^{-1/2}\mathbf{X} \tag{10.3.13}$$

The orthogonal characteristic vectors are not generally of unit length. Rather, the square of the length of the *i*th vector is given by:

$$\bar{\mathbf{x}}_i{}^T\bar{\mathbf{x}}_i = l_i \tag{10.3.14}$$

Therefore, to obtain a set of *orthogonal characteristic vectors of unit length* $\bar{\chi}_i$, the following transformation is necessary:

$$\boxed{\bar{\chi}_i = \frac{1}{\sqrt{l_i}}\bar{\mathbf{x}}_i} \tag{10.3.15}$$

Again, the matrix of orthogonal characteristic vectors of unit length $\bar{\chi}$ is related to the matrix of orthogonal characteristic vectors $\bar{\mathbf{X}}$ as follows:

$$\bar{\chi} = \bar{\mathbf{X}}\mathbf{L}^{-1/2} \tag{10.3.16}$$

where $\mathbf{L}^{-1/2}$ is the diagonal matrix:

$$\mathbf{L}^{-1/2} = \begin{bmatrix} 1/\sqrt{l_1} & 0 & 0 \\ 0 & 1/\sqrt{l_2} & 0 \\ 0 & 0 & 1/\sqrt{l_3} \end{bmatrix} \tag{10.3.17}$$

How can we use relations (10.3.12) and (10.3.15) to find straight-line reaction paths?

Suppose that, for a three-component system, we have found $\mathbf{x}_0$ by means of the equilibrium composition, as well as $\mathbf{x}_1$ by a technique to be discussed presently.

Then, relations (10.3.12) can be used to obtain the orthogonal characteristic vectors $\bar{\mathbf{x}}_0$ and $\bar{\mathbf{x}}_1$. The problem is then simply: Construct a vector orthogonal to both $\bar{\mathbf{x}}_0$ and $\bar{\mathbf{x}}_1$, transform it back to the nonorthogonal system, adjust its length — it will be the desired characteristic vector $\mathbf{x}_2$.

First construct a vector $\bar{\gamma}_1$ orthogonal to $\bar{\mathbf{x}}_0$. This is done easily by setting all elements of $\bar{\mathbf{x}}_0$ equal to zero, except two. These two elements must then be interchanged and one of them taken with the opposite sign. Then $\bar{\gamma}_1{}^T\bar{\mathbf{x}}_0 = 0$ so that both vectors are orthogonal. But $\bar{\gamma}_1$ is not in general orthogonal to $\bar{\mathbf{x}}_1$. Let us now adjust $\bar{\mathbf{x}}_1$ to unit length by means of (10.3.15). The projection of $\bar{\gamma}_1$ on $\bar{\chi}_1$ is a vector $\epsilon_{11}\bar{\chi}_1$. Thus:

$$\bar{\gamma}^T\bar{\chi}_1 = \epsilon_{11} \tag{10.3.18}$$

Vector subtraction of $\epsilon_{11}\bar{\chi}_1$ from $\bar{\gamma}_1$ gives a vector $\bar{\gamma}_2$ which is now perpendicular to both $\bar{\mathbf{x}}_0$ and $\bar{\mathbf{x}}_1$:

$$\bar{\gamma}_2 = \bar{\gamma}_1 - \epsilon_{11}\bar{\chi}_1 \tag{10.3.19}$$

The vector $\bar{\gamma}_2$ is then transformed back to the nonorthogonal system by the inverse of the transformation (10.3.12). The resulting vector will be the characteristic vector $\bar{\mathbf{x}}_2$ after its length has been adjusted by means of the second relation (10.2.21).

In a system with more than three components, the vector thus obtained would not be $\mathbf{x}_2$ since there would be little chance that $\bar{\gamma}_2$ would be orthogonal to the characteristic vectors yet to be determined. But the corresponding vector $\boldsymbol{\alpha}'_{x_2}(0)$ would provide a trial composition from which the true $\boldsymbol{\alpha}_{x_2}(0)$ might be determined experimentally by a converging process.

This determination is best understood by examining the situation in the case of a three-component system. The first experiment consists in obtaining data on the composition of the system at various times, as it approaches equilibrium, starting from some conveniently chosen composition (e.g., one of the pure components, say A_2). The reaction path shown in Fig. 10.2.1 is thus obtained. The portion of the reaction path near the equilibrium point tends to become tangent to the straight-line reaction path associated with the characteristic vector corresponding to the lowest characteristic root (in absolute value). Indeed, for sufficiently large values of time, if $\lambda_1 < \lambda_2$, $e^{-\lambda_1 t} \gg e^{-\lambda_2 t}$ and (10.2.23) approaches (10.2.25). The best straight line passing through the equilibrium point (a_2^*,a_3^*) and through a point $(\langle a_2\rangle,\langle a_3\rangle)$ determined by the average value of the points (a_2,a_3) on the reaction path near the equilibrium point, is given by the equation:

$$\frac{a_2 - a_2^*}{a_3 - a_3^*} = \frac{\langle a_2\rangle - a_2^*}{\langle a_3\rangle - a_3^*} \tag{10.3.20}$$

This straight line will intersect the side of the reaction triangle opposite to A_3 at a point of coordinates:

$$a_3 = 0$$

$$a_2 = a_2^* - a_3 \frac{\langle a_2 \rangle - a_2^*}{\langle a_3 \rangle - a_3^*} \qquad \text{[from (10.3.20)]} \qquad (10.3.21)$$

$$a_1 = 1 - a_2$$

which are the components of a vector

$$\boldsymbol{\alpha}_{x1}'(0) = \begin{bmatrix} 1 - a_2 \\ a_2 \\ 0 \end{bmatrix}$$

defining a new initial composition. This new initial composition is then used in a second experiment defining a second reaction path. The procedure described above is repeated until satisfactory agreement is obtained between initial composition and the next new initial composition calculated from (10.3.20). The last reaction path is the first straight-line reaction path.

The same procedure is applied to finding the next straight-line reaction path (or characteristic vector) in a system of more than three components. But it is important, as explained above, to start from a composition defined by a vector orthogonal to the first two orthogonal characteristic vectors. Otherwise, the reaction path, near equilibrium, will again be tangent to the straight-line reaction path already determined.

If the experimental data are reasonably accurate, the convergence in the determination of successive characteristic vectors appears to be satisfactorily rapid.

10.4 *Calculation of Rate Constants*

After experimental determination of the matrix $\mathbf{X}$ of the characteristic vectors, the rate constant matrix $\mathbf{K}$ can be calculated from (10.2.8). But this necessitates first a calculation of the inverse matrix $\mathbf{X}^{-1}$ and a determination of the characteristic roots.

The inversion of $\mathbf{X}$ can be carried out readily for the case of a network of reversible reactions. Indeed, since

$$\bar{\mathbf{x}}_i^T \mathbf{x}_i = 1 \qquad (10.4.1)$$

for any of the orthogonal characteristic vectors of unit length, a similar relation applies to the matrices of these vectors. Hence:

$$\bar{\mathbf{X}}^{-1} = \bar{\mathbf{X}}^{T} \tag{10.4.2}$$

From (10.3.16) and (10.3.13) it follows that:

$$\begin{aligned}\bar{\mathbf{X}}^{-1} &= (\bar{\mathbf{X}}\mathbf{L}^{-1/2})^{-1} = \mathbf{L}^{1/2}\bar{\mathbf{X}}^{-1} = \mathbf{L}^{1/2}(\mathbf{D}^{-1/2}\mathbf{X})^{-1} = \mathbf{L}^{1/2}\mathbf{X}^{-1}\mathbf{D}^{1/2}\\ \bar{\mathbf{X}}^{T} &= (\bar{\mathbf{X}}\mathbf{L}^{-1/2})^{T} = \mathbf{L}^{-1/2}\bar{\mathbf{X}}^{T} = \mathbf{L}^{-1/2}(\mathbf{D}^{-1/2}\mathbf{X})^{T} = \mathbf{L}^{-1/2}\mathbf{X}^{T}\mathbf{D}^{-1/2}\end{aligned} \tag{10.4.3}$$

Therefore:

$$\mathbf{L}^{1/2}\mathbf{X}^{-1}\mathbf{D}^{1/2} = \mathbf{L}^{-1/2}\mathbf{X}^{T}\mathbf{D}^{-1/2} \tag{10.4.4}$$

Thus:

$$\boxed{\mathbf{X}^{-1} = \mathbf{L}^{-1}\mathbf{X}\mathbf{D}^{-1}} \tag{10.4.5}$$

The matrix $\mathbf{L}^{-1}$ is obtained readily by noting that, according to (10.3.14):

$$\bar{\mathbf{X}}^{T}\bar{\mathbf{X}} = \mathbf{L} \tag{10.4.6}$$

Because of (10.3.13), (10.4.6) gives:

$$(\mathbf{D}^{-1/2}\mathbf{X})^{T}(\mathbf{D}^{-1/2}\mathbf{X}) = \mathbf{X}^{T}\mathbf{D}^{-1/2}\mathbf{D}^{-1/2}\mathbf{X} = \mathbf{L}$$

Therefore:

$$\boxed{\mathbf{L} = \mathbf{X}^{T}\mathbf{D}^{-1}\mathbf{X}} \tag{10.4.7}$$

and since $\mathbf{L}$ is a diagonal matrix, its inverse is obtained at once.

Formulae (10.4.5) and (10.4.7) afford an easy way to calculate $\mathbf{X}^{-1}$. The remaining problem is to evaluate $\mathbf{\Lambda}$.

But with $\mathbf{X}^{-1}$, it is easy to transform $\boldsymbol{\alpha}$ into $\boldsymbol{\beta}$, by means of (10.2.11), which can be rewritten as:

$$\boxed{\boldsymbol{\beta} = \mathbf{X}^{-1}\boldsymbol{\alpha}} \tag{10.4.8}$$

The components b_i of the vector $\boldsymbol{\beta}$ can thus be obtained for any reaction path determined experimentally in the search for characteristic vectors. Because of the relations (10.2.22) a plot of ln b_i versus time gives a straight line of slope $-\lambda_i$.

If, however, only relative values of the rate constants are required, then only relative values of the characteristic roots need to be found. These can be obtained by noting that from the last two relations in (10.2.22):

$$\boxed{\ln b_1 = \text{const} + \frac{\lambda_1}{\lambda_2} \ln b_2} \tag{10.4.9}$$

Thus a plot of ln b_1 versus ln b_2 yields a straight line of slope λ_1/λ_2. Similarly, in a system of more than three components, all roots λ_i $(i = 1,2, \ldots i, \ldots n - 2)$ can be determined relative to λ_{n-1}. A matrix $\boldsymbol{\Lambda}'$ proportional to $\boldsymbol{\Lambda}$ can be built and a matrix $\mathbf{K}'$ proportional to $\mathbf{K}$ calculated. Finally, all rate constants relative to one of them taken as unity can be determined.

10.5 *Illustration: Isomerization of Butenes on Alumina*

This triangular reaction network, discovered by Haag and Pines and further studied by Lago and Haag, provides a recent but already classical example of a successful analysis of a network of first-order reversible reactions. The network is:

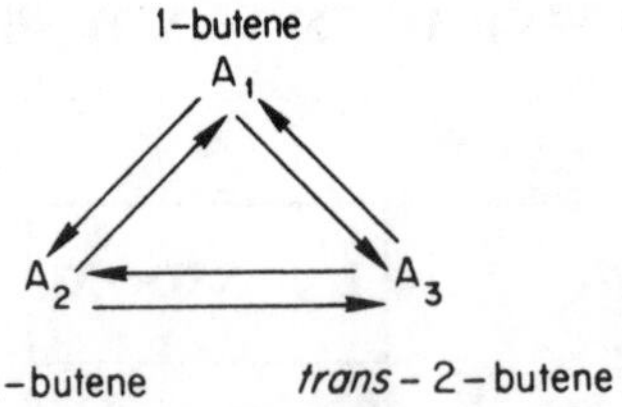

First of all, the equilibrium composition of the system, at a selected temperature, is determined experimentally. This gives the equilibrium vector $\boldsymbol{\alpha}^*$ identical to the zeroth characteristic vector $\mathbf{x}_0$:

$$\boldsymbol{\alpha}^* = \mathbf{x}_0 = \begin{bmatrix} 0.1436 \\ 0.3213 \\ 0.5351 \end{bmatrix} \tag{10.5.1}$$

Then, a reaction path, starting from an arbitrarily chosen composition, namely pure *cis*-butene, is determined experimentally. Since only relative values of the rate constant are desired, the value of the time is immaterial and the various time intervals at which the composition of the system is measured are simply denoted as t_0, t_1, t_2, etc. These compositions are collected in Table 10.5.1.

The reaction path would be similar to those plotted in Fig. 10.2.1. A straight line passing approximately through the last segment of the curve near equilibrium intercepts on the side A_1A_2 a point determining a new composition for the second trial:

$$\boldsymbol{\alpha}(0) = \begin{bmatrix} 0.240 \\ 0.760 \\ 0 \end{bmatrix} \tag{10.5.2}$$

Data not too far from the equilibrium point for a second reaction path are collected in Table 10.5.2. With the average values noted, a new initial composition

$$\boldsymbol{\alpha}'(0) = \begin{bmatrix} 0.3286 \\ 0.6714 \\ 0.0000 \end{bmatrix} \tag{10.5.3}$$

is calculated by means of Eqs. (10.3.21). The next initial composition used is

Table 10.5.1

THE FIRST EXPERIMENTAL CURVED REACTION PATH

	MOLE FRACTIONS	
	cis-2-butene	*trans*-2-butene
t_0	1.0000	0.0000
t_1	0.9191	0.0422
t_2	0.8897	0.0560
t_3	0.8477	0.0820
t_4	0.8177	0.0969
t_5	0.6603	0.2001
t_6	0.6487	0.2102
t_7	0.6354	0.2178
t_8	0.5230	0.3150

Table 10.5.2

SECOND EXPERIMENTAL REACTION PATH: DATA NOT FAR FROM EQUILIBRIUM

	Mole Fractions	
	cis-2-butene	*trans*-2-butene
	0.3604	0.4775
	0.3769	0.4455
	0.3595	0.4741
	0.3622	0.4724
	0.3671	0.4639
	0.3441	0.4955
	0.3471	0.4992
	0.3464	0.4965
	0.3431	0.5027
	0.3451	0.5028
	0.3408	0.5067
	0.3416	0.5052
average	0.35286	0.48683

Table 10.5.3

THIRD EXPERIMENTAL REACTION PATH: DATA NOT FAR FROM EQUILIBRIUM

	Mole Fractions	
	cis-2-butene	*trans*-2-butene
	0.4606	0.3105
	0.4738	0.2900
	0.4118	0.3894
	0.3915	0.4190
	0.3678	0.4571
	0.3815	0.4384
	0.3478	0.4965
	0.3589	0.4767
	0.3423	0.5000
	0.3395	0.5021
	0.3324	0.5167
	0.3290	0.5159
	0.3314	0.5152
average	0.37448	0.44827

as close to (10.5.3) as convenient. It is:

$$\boldsymbol{\alpha}''(0) = \begin{bmatrix} 0.3258 \\ 0.6742 \\ 0.0000 \end{bmatrix} \tag{10.5.4}$$

With the starting composition (10.5.4) a new run is performed. Corresponding data not far from equilibrium are collected in Table 10.5.3. Again from the average value of the points and from (10.3.21) an initial composition:

$$\boldsymbol{\alpha}'''(0) = \begin{bmatrix} 0.3510 \\ 0.6490 \\ 0.0000 \end{bmatrix} \tag{10.5.5}$$

is calculated. Once more, this composition is duplicated as closely as possible and a starting composition:

$$\boldsymbol{\alpha}''''(0) = \begin{bmatrix} 0.3551 \\ 0.6449 \\ 0.0000 \end{bmatrix} \tag{10.5.6}$$

is used for a last run shown in Table 10.5.4. The data of this run give a starting composition

$$\boldsymbol{\alpha}_{x_1}(0) = \begin{bmatrix} 0.3492 \\ 0.6508 \\ 0.0000 \end{bmatrix} \tag{10.5.7}$$

which is almost identical to (10.5.6) and is therefore adopted as the starting composition of the true first straight-line reaction path. Then, from (10.2.21), we get, using (10.5.1) and (10.5.7),

$$\mathbf{x}_1 = \begin{bmatrix} 0.3492 \\ 0.6508 \\ 0.0000 \end{bmatrix} - \begin{bmatrix} 0.1436 \\ 0.3213 \\ 0.5351 \end{bmatrix} = \begin{bmatrix} 0.2056 \\ 0.3295 \\ -0.5351 \end{bmatrix} \tag{10.5.8}$$

For this problem, the experimental work is now completed. The rest of

Table 10.5.4

FOURTH AND LAST EXPERIMENTAL REACTION PATH: DATA NOT FAR FROM EQUILIBRIUM

MOLE FRACTIONS

	cis-2-butene	*trans*-2-butene
	0.5689	0.1337
	0.5642	0.1447
	0.5386	0.1814
	0.5202	0.2139
	0.5043	0.2380
	0.4758	0.2798
	0.4579	0.3110
	0.4281	0.3644
	0.4031	0.4031
	0.3618	0.4668
average	0.4823	0.2737

the work is calculation.

First determine the auxiliary matrices $\mathbf{D}^{1/2}$, $\mathbf{D}^{-1/2}$ and $\mathbf{D}^{-1}$ by means of (10.5.1). Off-diagonal zeroes of diagonal matrices need not be written:

$$\mathbf{D}^{-1} = \begin{bmatrix} 6.9638 & & \\ & 3.1123 & \\ & & 1.8688 \end{bmatrix} \tag{10.5.9}$$

$$\mathbf{D}^{1/2} = \begin{bmatrix} 0.3789 & & \\ & 0.5668 & \\ & & 0.7315 \end{bmatrix} \tag{10.5.10}$$

$$\mathbf{D}^{-1/2} = \begin{bmatrix} 2.6389 & & \\ & 1.7642 & \\ & & 1.3670 \end{bmatrix} \tag{10.5.11}$$

Next calculate the vectors $\bar{\mathbf{x}}_0$ and $\bar{\mathbf{x}}_1$ using (10.3.12):

$$\bar{\mathbf{x}}_0 = \begin{bmatrix} 0.3789 \\ 0.5668 \\ 0.7315 \end{bmatrix} \qquad \bar{\mathbf{x}}_1 = \begin{bmatrix} 0.5426 \\ 0.5812 \\ -0.7315 \end{bmatrix} \tag{10.5.12}$$

The square of the length of $\bar{\mathbf{x}}_1$ is:

$$\bar{\mathbf{x}}_1^T\bar{\mathbf{x}}_1 = 0.5426^2 + 0.5812^2 + 0.7315^2 = 1.1673 = l_1$$

Therefore, according to (10.3.15), we obtain $\bar{\mathbf{x}}_1$ by means of:

$$\bar{\chi}_1 = \frac{1}{\sqrt{l_1}}\bar{\mathbf{x}}_1 = \begin{bmatrix} 0.5022 \\ 0.5380 \\ -0.6771 \end{bmatrix} \qquad (10.5.13)$$

Construct $\bar{\gamma}_1$ orthogonal to $\bar{\mathbf{x}}_0$:

$$\bar{\gamma}_1 = \begin{bmatrix} -0.5668 \\ 0.3789 \\ 0.0000 \end{bmatrix} \qquad (10.5.14)$$

Since

$$\bar{\chi}_1^T\bar{\gamma}_1 = \epsilon_{11} = -0.080799$$

the vector $\bar{\gamma}_2$

$$\bar{\gamma}_2 = \bar{\gamma}_1 + 0.080799\,\bar{\chi}_1 = \begin{bmatrix} -0.5262 \\ 0.4224 \\ -0.0547 \end{bmatrix} \qquad (10.5.15)$$

is also orthogonal to $\bar{\mathbf{x}}_2$ as shown in connection with Eq. (10.3.19).

Next transform $\bar{\gamma}_2$ back to nonorthogonal system:

$$\bar{\gamma}_2 = \mathbf{D}^{1/2}\bar{\gamma}_2 = \begin{bmatrix} -0.1994 \\ 0.2394 \\ -0.0400 \end{bmatrix} \qquad (10.5.16)$$

Finally, adjust the length of $\bar{\gamma}_2$ by multiplication by (0.1436/0.1994) so that by addition to $\bar{\mathbf{x}}_0$ it gives a vector $\boldsymbol{\alpha}_{x_2}(0)$ the first component of which is zero. This gives $\mathbf{x}_2$:

$$\mathbf{x}_2 = \begin{bmatrix} -0.1436 \\ 0.1724 \\ -0.0288 \end{bmatrix} \qquad (10.5.17)$$

With (10.5.1), (10.5.8) and (10.5.17), from the matrix $\mathbf{X}$:

$$\mathbf{X} = \begin{bmatrix} 0.1436 & 0.2056 & -0.1436 \\ 0.3213 & 0.3295 & 0.1724 \\ 0.5351 & -0.5351 & -0.0288 \end{bmatrix} \tag{10.5.18}$$

Then, by means of (10.4.7), calculate $\mathbf{L}$, using (10.5.9) and (10.5.18):

$$\mathbf{L} = \begin{bmatrix} 0.1436 & 0.3213 & 0.5351 \\ 0.2056 & 0.3295 & -0.5351 \\ -0.1436 & 0.1724 & -0.0288 \end{bmatrix} \begin{bmatrix} 6.9638 & & \\ & 3.1123 & \\ & & 1.8688 \end{bmatrix}$$

$$\begin{bmatrix} 0.1436 & 0.2056 & -0.1436 \\ 0.3213 & 0.3295 & 0.1724 \\ 0.5351 & -0.5351 & -0.0288 \end{bmatrix} = \begin{bmatrix} 1.0000 & & \\ & 1.1674 & \\ & & 0.2371 \end{bmatrix}$$

Therefore:

$$\mathbf{L}^{-1} = \begin{bmatrix} 1.0000 & & \\ & 0.8566 & \\ & & 4.2077 \end{bmatrix} \tag{10.5.19}$$

Next, from (8.4.5), using (8.5.19), (8.5.18) and (8.5.9), we get:

$$\mathbf{X}^{-1} = \begin{bmatrix} 1.0000 & 1.0000 & 1.0000 \\ 1.2265 & 0.8784 & -0.8566 \\ -4.2077 & 2.2579 & -0.2264 \end{bmatrix} \tag{10.5.20}$$

We are now ready to calculate $\mathbf{\Lambda}$. Indeed the values of $\boldsymbol{\alpha}$ in Table 10.5.1 can be transformed into $\boldsymbol{\beta}$ by means of (10.4.8). The corresponding values of b_1 and b_2 are collected in Table 10.5.5. Plotting ln b_1 versus ln b_2, we obtain a straight line of slope $\lambda_1/\lambda_2 = 0.4769$ as shown by (10.4.9).

This is sufficient to write down a matrix $\mathbf{\Lambda}'$ proportional to $\mathbf{\Lambda}$:

$$\mathbf{\Lambda}' = \begin{bmatrix} 0.0000 & & \\ & -0.4769 & \\ & & -1.0000 \end{bmatrix} \tag{10.5.21}$$

The product of matrices (10.5.20), (10.5.21) and (10.5.18) gives a matrix $\mathbf{K}'$ proportional to $\mathbf{K}$:

$$\mathbf{K}' = \begin{bmatrix} -0.7245 & 0.2381 & 0.0515 \\ 0.5327 & -0.5273 & 0.1736 \\ 0.1918 & 0.2892 & -0.2251 \end{bmatrix} \tag{10.5.22}$$

Table 10.5.5

COMPONENTS OF THE VECTOR $\boldsymbol{\beta}$ REPRESENTING THE FIRST REACTION PATH

$(b_0 = 1)$

	b_1	b_2
t_0	0.8784	2.2579
t_1	0.8187	1.9028
t_2	0.8001	1.7677
t_3	0.7607	1.5995
t_4	0.7400	1.4650
t_5	0.5798	0.8582
t_6	0.5629	0.8233
t_7	0.5517	0.7676
t_8	0.3883	0.4279

Comparison of (10.5.22) with (10.2.2) shows that if k_{31} is taken as unity, we have determined all rate constants:

$$k_{12} = 10.344 \qquad k_{13} = 3.724 \qquad k_{23} = 5.616$$
$$k_{21} = 4.623 \qquad k_{31} = 1.000 \qquad k_{32} = 3.371$$

10.6 Kinetic Response of Reaction Networks

The formulation of Wei and Prater offers many advantages. Its limitations are not as confining as suggested by the single application discussed in this chapter, namely, the analysis of first-order reversible reactions. The method can be extended readily to networks of first-order reactions that are not all reversible. Furthermore, as discussed in Chapter 5, rate equations for the coupled sequences of a network are very frequently of a type that can still be handled by the analysis. They are commonly, though not always, of the form:

$$r = \varphi k c_{A_i} \tag{10.6.1}$$

where φ is a function, usually appearing in the denominator, *the same for all coupled sequences of the network.* Then, the system of differential equations can be written in matrix notation as:

$$\frac{d\boldsymbol{\alpha}}{dt} = \varphi\mathbf{K}\boldsymbol{\alpha} \tag{10.6.2}$$

and (10.6.2) can be treated by the method of Wei and Prater.

The method of Prater and Wei also affords an interesting way to familiarize oneself with the kinetic response of reaction networks and their experimental analysis. In the case of a three-component system, for instance, this familiarity can be achieved in the following way.

Let us imagine that we know the rate constants, and therefore $\mathbf{K}$. Then we can find the characteristic vectors and roots. Select any vector, e.g.,

$$\mathbf{x}' = \begin{bmatrix} 1 \\ -1 \\ 0 \end{bmatrix}$$

The operation $\mathbf{K}\mathbf{x}'$ yields a vector $\mathbf{y}'$ that can be written as

$$-\lambda' \begin{bmatrix} 1 \\ a - 1 \\ -a \end{bmatrix} = -\lambda'\mathbf{x}''$$

Repeating the operation on $\mathbf{x}''$ we get in the same manner $\mathbf{K}\mathbf{x}'' = \mathbf{y}''$ and $\mathbf{y}'' = \lambda''\mathbf{x}'''$. We will soon converge, i.e., we will find a vector $\mathbf{x}_2$ such that

$$\mathbf{K}\mathbf{x}_2 = -\lambda_2\mathbf{x}_2 \tag{10.6.3}$$

Clearly $\mathbf{x}_2$ is a characteristic vector and it can be shown that λ_2 is the largest root in absolute value. Then $\mathbf{x}_1$ can be found by the orthogonality relations and λ_1 from a relation analogous to (10.6.3). In a multicomponent system with $(n - 1)$ nonzero roots $|\lambda_1| < |\lambda_2| \ldots < |\lambda_{n-2}| < |\lambda_{n-1}|$, the value of λ_{n-1} is first found as just shown. Then $|\lambda_{n-2}|$ can be determined by the same method applied to the matrix $\mathbf{K} + |\lambda_{n-1}|\, I$ instead of $\mathbf{K}$, and so on. Having determined characteristic roots, we can construct the diagonal matrix:

$$e^{\lambda t} = \begin{bmatrix} 1 & 0 & 0 \\ 0 & e^{-\lambda_1 t} & 0 \\ 0 & 0 & e^{-\lambda_2 t} \end{bmatrix} \tag{10.6.4}$$

and perform the necessary "experiments" in order to determine the rate constants if we now pretend that we do not know them. These "experiments" are performed by means of the relation:

$$\boxed{\boldsymbol{\alpha}(t) = Xe^{\lambda t}\,\mathbf{X}^{-1}\boldsymbol{\alpha}(0)} \tag{10.6.5}$$

which is easily obtained. Indeed,

$$\boldsymbol{\alpha}(t) = \mathbf{X}\boldsymbol{\beta}(t) \tag{10.2.11}$$

$$\boldsymbol{\beta}(t) = e^{\lambda t}\boldsymbol{\beta}(0) \tag{10.2.22}$$

$$\boldsymbol{\beta}(0) = \mathbf{X}^{-1}\boldsymbol{\alpha}(0) \tag{10.4.8}$$

These "experiments" that simulate work in the laboratory illustrate the essential feature of the method when performed in a real situation: the experimental runs are used as an analog computer to determine the rate constants more rapidly and more safely.

Finally, an important theoretical result emerges from the preceding treatment. The formulation discussed in this chapter shows clearly that purely kinetic oscillations are impossible in the case of a network of reversible first-order reactions. This was first demonstrated by Jost (1947) and is a consequence of the principle of microscopic reversibility.

Indeed, immediately following Eq. (10.2.3) it was *assumed* that the matrix $\mathbf{K}$ for an n-component system possessed n characteristic nonpositive roots. This assumption is justified by the fact that the similarity transformation $\mathbf{D}^{-1/2}\mathbf{K}\mathbf{D}^{1/2}$ gives a symmetric matrix $\bar{\mathbf{K}}$, as was shown in Section 10.3 to be the case as a result of the principle of microscopic reversibility. But it is known from matrix algebra that a symmetric $(n \times n)$ matrix possesses n characteristic *nonpositive real* roots. Since, furthermore, $\bar{\mathbf{K}}$ has the same roots as $\mathbf{K}$, as was mentioned in Section 10.3, the assumption made is justified. Therefore any composition vector $\boldsymbol{\alpha}$ will go to its equilibrium value without oscillating about the latter, as shown clearly by Eq. (10.2.23).

The absence of purely kinetic oscillations in the case of a network of first-order reactions does not exclude the possibility of periodic phenomena in chemically reacting systems. Indeed many such phenomena are known and their study has exerted a particular fascination on chemical kineticists from the early days of physical chemistry. With the rapid development of kinetics in biochemistry and biophysics, the interest in periodic phenomena is rapidly increasing, especially in connection with the study of photoperiodism in plants and animals, of rhythmic processes in living organisms and of "clock phenomena" in general.

The possibility of oscillations in chemical kinetics was first suggested as a possible application of a mathematical model representing the growth of populations devouring each other. But the rate equations of the network must then be of the autocatalytic type (see Chapter 6) and this is a rather confining limitation. More recently, the possibility of kinetic oscillations in a much more realistic network of reactions has been demonstrated experimentally and theoretically in enzyme systems involving feedback loops.

In such studies, the use of analog and digital computers pioneered by Britton Chance, plays a decisive role. Thus, networks of enzyme reactions in living cells, consisting of up to one hundred reactions, are now under active scrutiny.

As a particularly simple example of a network exhibiting oscillatory behavior, consider the network of Higgins (1964) represented schematically as follows:

$$A \rightarrow B \qquad (1)$$

$$B \xrightarrow{E_1^*} C \qquad (2)$$

$$C + E_1 \rightleftarrows E_1^* \qquad (3)$$

$$C \rightarrow D \qquad (4)$$

Reactions (1) and (4) are catalyzed by enzymes. The essential feature of the network consists in the fact that reaction (2) is catalyzed by an enzyme E_1^* which is activated by reaction (3) of an inactive form of the enzyme, E_1, with the reaction product of (2). Higgins has demonstrated that for certain values of the rate constants and of the concentrations, this network of four reactions will oscillate.

Even for the analysis of such a simple network, the use of a computer is mandatory. In certain cases, when it is possible to use a well-stirred reactor (Chapter 1), the analysis of the network is simplified greatly, since the system of differential equations is replaced by a system of algebraic equations. In any case, the analysis of reaction networks remains the most elaborate problem facing the chemical kineticist in pure or applied work.

Problem 10.6.1

Consider the network of first-order reactions

$$C \underset{1}{\overset{1}{\leftrightarrows}} A \underset{10}{\overset{10}{\rightleftarrows}} B$$

The rate constants are the figures above and below the arrows, in arbitrary units. Determine $\mathbf{X}$, $e^{\lambda t}$ and $\mathbf{X}^{-1}$. Perform "experiments" with the help of (10.6.5), so as to obtain the rate constants now assumed to be unknown. Examples are shown in Fig. 10.6.1.

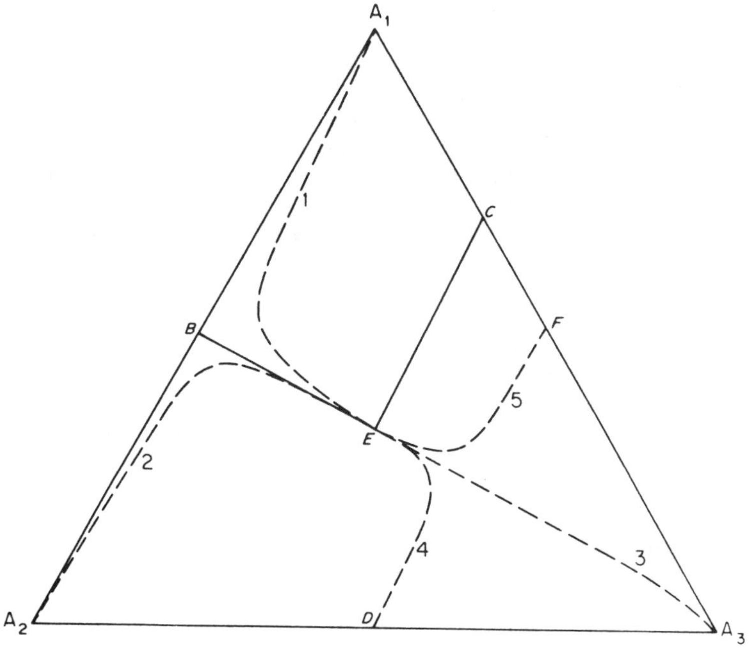

Fig. 10.6.1 Kinetic response for the reaction network $A_3 \underset{1}{\overset{1}{\leftrightarrows}} A_1 \underset{10}{\overset{10}{\rightleftarrows}} A_2$. Solid lines BE and CE represent the straight-line reaction paths from initial compositions B ($a_1 = 0.486$, $a_2 = 0.514$, $a_3 = 0$) and C ($a_1 = 0.685$, $a_2 = 0$, $a_3 = 0.315$), respectively, toward equilibrium E. The reaction paths starting from each of the pure components are shown (trajectories 1, 2 and 3). The reaction paths from initial compositions of D ($a_1 = 0$, $a_2 = 0.5$, $a_3 = 0.5$) and F ($a_1 = 0.5$, $a_2 = 0$, $a_3 = 0.5$) are shown (trajectories 4 and 5, respectively).

BIBLIOGRAPHY

Analytical solutions of reaction networks can be found in *Kinetics and Mechanism* by A. A. Frost and R. G. Pearson, John Wiley and Sons, Inc., New York, 1961, and in *Cinétique Chimique Appliquée* by Jungers, *et al.*, Technip, Paris, 1956. More solutions are found in a review paper by R. Zahradnik, *Chem. Listy*, **53**, 56 (1959). Numerical techniques of solution using digital computers have been developed by T. I. Peterson, *Chem. Eng. Progr. Symp. Ser.* **56**, No. 31, 111 (1960); *Chem. Eng. Sci.*, **17**, 293 (1962); G. W. Booth and T. I. Peterson, *A.I.Ch.E. Computer Program Manual* No. 3, American Institute of Chemical Engineers, New York, 1960.

10.1 Some interesting applications of Neiman's kinetic tracer technique are discussed by C. F. Cullis, A. Fish and D. K. Trimm, *Ninth Symposium (International) on Combustion*, Academic Press Inc., New York, 1960, p. 167.

10.2
10.5 The material in these sections has been treated in great detail by the originators of the method, James Wei and Charles D. Prater, *Advances in Catalysis,* Academic Press Inc., New York, 1962, Vol. 13, p. 204. Readers not familiar with definitions and fundamental operations of matrices will find this material in many textbooks, e.g., Neal R. Amundson, *Mathematical Methods in Chemical Engineering: Matrices and Their Application,* Prentice-Hall, Inc., Englewood Cliffs, N.J., 1966.

10.6 An interesting historical survey of periodic phenomena in chemical kinetics is found in a book by E. S. Hedges and J. E. Myers, *The Problem of Physico-Chemical Periodicity*, Edward Arnold & Co., London, 1926. The general proof that the characteristic roots of the rate-constant matrix are nonpositive real numbers was first given by W. Jost, *Z. Naturforsch.*, **2a**, 159 (1947). A fictitious "experimental" program of the kind discussed here is given in detail by Wei and Prater for a system of four components. Kinetic oscillations in enzyme systems were first demonstrated by J. Higgins, *Proc. Nat. Ac. Sci.*, **51**, 989 (1964).

APPENDIX

Glossary Of Terms Used In Chemical Kinetics

Activated complex	See transition state.
Activation barrier	For an exothermic elementary step, it is the difference in internal energy between transition state and reactants.
Activation energy	For an elementary step, it is the difference in internal energy between transition state and reactants.
Activation enthalpy	For an elementary step, it is the difference in enthalpy between transition state and reactants.
Activation entropy	For an elementary step, it is the difference in standard entropy between transition state and reactants. The standard state must, of course, be defined in each situation.
Active center	Any reactive intermediate intervening between reactants and products in a single reaction, though not appearing in the stoichiometric equation for reaction.
Active sites	Groups at the surface of a solid or enzyme, responsible for their catalytic activity.
Arrhenius activation energy	The temperature coefficient E in the Arrhenius expression for a rate constant $k = A \exp(-E/RT)$ where A is temperature-independent and called the pre-exponential factor or frequency factor.
Autocatalytic reaction	A reaction which produces its own catalyst.
Branched-chain reaction	A chain reaction in which there is a net gain in the number of active centers as the closed sequence repeats itself.
Branching step	A step in a chain reaction in which there is a net gain in the number of active centers.
Cage effect	Active centers produced in a medium of high density by decomposition of an initiator may not succeed in diffusing away from each other before they recombine.
Catalyst	A source of active centers regenerated at the end of a closed reaction sequence.
Catalytic reaction	A closed sequence which owes its propagation to a substance called a catalyst.
Chain carriers	The active centers in a chain reaction.

Chain length	See *Kinetic chain length.*
Chain reaction	A closed sequence which owes its propagation to some external agency (light, ionizing radiation) or to a molecular species susceptible of generating active centers by its own destruction.
Closed sequence	A reaction sequence in which the active center consumed in the first step is regenerated in the last one.
Coupling	In a network, independent reactions frequently proceed through identical or similar active centers. In this case, their reaction sequences are kinetically coupled.
Degenerate branching	A branching step which consists of the reaction of a relatively stable intermediate.
Delayed branching	See *Degenerate branching.*
Elementary step	The irreducible act of reaction in which reactants are transformed into products directly, i.e., without passing through an intermediate that is susceptible of isolation.
Extent of reaction	See *Single reaction.*
Frequency factor	See *Arrhenius activation energy.*
Gel effect	This occurs when the rate of termination is controlled by diffusion of active centers toward each other.
Induction period	The time during which the extent of an autocatalytic reaction remains below experimental detection.
Inhibition	Decrease in rate occasioned by a substance (inhibitor, poison) which may be produced by the reaction itself or may be a foreign substance.
Initiation	A step in which active centers are produced by the decomposition (thermal, photolytic, radiolytic) of stable species (initiators).
Irreversible	A reaction is irreversible if the rate of the reverse reaction can be neglected as compared to the rate in the forward direction.
Kinetic chain length	If it is a sufficiently large number, it is the ratio of the rates of propagation and initiation in a chain reaction. This definition applies as a result of the *Long-chain approximation.*
Kinetic mechanism	See *Mechanism.*
Long-chain approximation	See *Kinetic chain length.*
Mechanism	A vague term, related to Latin "machina," used loosely to describe a reaction network, or a reaction sequence, or the stereochemistry of an elementary step. Sometimes called a "model." If based on kinetic arguments, it is occasionally called a "kinetic mechanism."
Model	See *Mechanism.*

Molecularity	An elementary step is uni-, bi-, or ter-molecular if it involves respectively one, two or three reactants.
Network	When several single reactions take place in a system, these parallel and consecutive reactions constitute a network.
Open sequence	A reaction sequence in which the active center produced in the first step is consumed in the last one.
Order	A reaction is of order *n* with respect to a given component if its rate is proportional to the concentration of that component, to the *n*th power.
Penetration effect	In any heterogeneous reaction, reactants have to diffuse from one phase to an interface or to another phase: If the resulting concentration gradients are not negligible, the rate of the chemical reaction will be obscured by this penetration effect.
Pre-exponential factor	See *Arrhenius activation energy*.
Propagation	The steps of the closed sequence in a chain reaction.
Rate constant	The constant of proportionality that does not depend on composition, in the expression for the rate of a reaction. Also called "specific rate constant."
Rate-determining step	If, in a reaction sequence, consisting of *n* steps, $(n-1)$ steps are reversible and if the rate of each one is very much larger in either direction than the rate of the *n*th step, the latter is said to be rate-determining. It is also sometimes called the slow step. The rate-determining step need not be reversible.
Reaction sequence	A succession of elementary steps, the sum of which reproduces the stoichiometric equation of a single reaction.
Relaxation time	Time during which the concentration of a given species decays to a fraction $(1/e)$ of its original value.
Reversible	A reaction is reversible if the rates in the forward and reverse direction are of the same order of magnitude.
Sequence	See *Reaction sequence*.
Single reaction	A reaction the advancement of which may be characterized by a single parameter called the extent of reaction.
Slow step	See *Rate-determining step*.
Steady-state approximation	Its mathematical expression is that the time rate of change of the concentration of all active centers in a reaction sequence is equal to zero.
Step	See *Elementary step*.
Termination	A step which destroys active centers in a chain reaction.
Transition state	The configuration of highest potential energy along the path of lowest energy between reactants and products. Synonymous with *activated complex*.

Uniform A surface is said to be uniform when all its active sites form active centers of identical thermodynamic and kinetic properties.

Wall effect The rate of many homogeneous reactions is affected by the presence and state of reactor walls.

INDEX